QUARKS
TO CULTURE

——

TYLER VOLK

QUARKS TO CULTURE

How We Came to Be

COLUMBIA UNIVERSITY PRESS
NEW YORK

Columbia University Press
Publishers Since 1893
New York   Chichester, West Sussex
cup.columbia.edu

Library of Congress Cataloging-in-Publication Data

Names: Volk, Tyler.
Title: Quarks to culture : how we came to be / Tyler Volk.
Description: New York : Columbia University Press, [2017] | Includes
bibliographical references and index.
Identifiers: LCCN 2016045624 | ISBN 9780231179607 (cloth : alk. paper) |
ISBN 9780231544139 (e-book)
Subjects: LCSH: Life—Origin.
Classification: LCC QH325 .V588 2017 | DDC 576.8/3—dc23
LC record available at https://lccn.loc.gov/2016045624

Columbia University Press books are printed on permanent
and durable acid-free paper.
Printed in the United States of America

Cover design: Faceout Studio

To Amelia
and everyone in
my metagroup of love

CONTENTS

PART 3. DYNAMICAL REALMS AND THEMES

A Graphic Concept Summary can be found between pages 78 and 79

PREFACE

This work is a new study of patterns, systems, things. My original impetus was to write a direct follow-up to an earlier book, *Metapatterns Across Space, Time, and Mind*.[1] In that work, I had permitted myself the freedom of a high-flying bird, seeking themes and similarities across fields of natural science and even social sciences and humanities.

Metapatterns are scale-bridging patterns, with ties to geometric arrangements of parts in wholes. These patterns often recur in biology and cultural systems because of properties they have that can offer functional advantages. Examples include spheres, tubes, borders, organizational centers, and cycles.

Progress, I felt, would emphasize patterns that come about from evolutionary processes versus those that do not. That idea did turn out to be a key motif here. But the conceptual zygote from which this new work grew truly took off with a simple insight about the body. A question I had walked up to in the earlier work turned into an expansive doorway.

The metapattern called "layers" concerns the widely found pattern of a nested hierarchy (or, as I prefer, holarchy) of parts and wholes in the most general sense. In my earlier work, I had guessed there might be fundamental layers (levels) but had left hanging the question about how to find them. Now I realized that the question could be put into very concrete form by thinking from a particular angle about things in the human body *and also about time*.

As you go down into the body, you go back in time: from the body inward to cells, to molecules, and then to atoms. Passing from life to physics, each first type in this series of nested things came into existence earlier.

That's interesting, I said to myself. *Would it be possible to discern fundamental levels?* Let's reverse the list, going from ancient to modern time. Can one start at the simplest things of physics and ratchet along a course in time that simultaneously progresses outward in scale? And perhaps during this tally, let's not halt at our bodies as a terminal level but continue the logic on up to larger patterns that we as bodies and minds participate in, such as the social systems of complex culture.

Jacob Bronowski, scientist, master explainer, and bridge between the natural sciences and the humanities, developed in *The Ascent of Man* a concept that he named "stratified stability." I still recall evenings in the mid-1970s when I was between architecture school and future graduate work in earth systems science and ripe for hearing about deep patterns in nature. A group of friends and I lounged weekly, glued to the thirteen televised episodes covering the ideas in Bronowski's book. The words from one episode still ring for me: "Nature works in steps. The atoms form molecules. . . . [T]he cells make up . . . the simple animals. . . . [S]table units that compose one level or stratum are the raw material for . . . higher configurations . . . a ladder from simple to complex by steps, each of which is stable in itself. . . . I call [this] *Stratified Stability*."[2]

Bronowski did not attempt to count the strata (and he died young, at age sixty-six in 1974, still in the fullness of his powers). In addition, he applied his concept in ways both similar and dissimilar to my pursuit. For instance, in details not quoted here, it is clear that he did not distinguish, as I will, between the arrival of new configurations *within* a level and the transition *to* new levels.[3] And I want a term that emphasizes a *repeated process* in the forming of fundamental levels. Therefore, for better or worse, I have coined an active word: *combogenesis*.

Combogenesis is the combination and integration of things from a prior level to make a new level of things. As described in this book, a sequence of such events of combogenesis has over time produced an increasingly expansive nestedness of things from the simplest particles of physics, such as quarks, to the geopolitical state in human culture.

To be sure, many have remarked on and explored the attribute of nestedness. After chemists found the atom, physicists discovered it had parts,

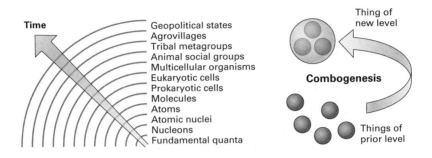

FIGURE P.1

Twelve levels of the grand sequence, shown as concentric rings. Each level was created from an event of combogenesis, with a generic event shown on the right. Combogenesis repeated as a rhythm that went from the fundamental quanta (quarks, etc.) to geopolitical states by a series of levels whose initial things combined and integrated things from previous levels. (On the right, the combined components are shown as a trio inside the thing or system at the upper, new level only for simplicity's sake.)

and one of those parts, the nucleus, had parts, and those parts contained quarks. The search for deeper layers of the onion continues apace.

The book *The Major Transitions of Evolution* by Maynard Smith and Eörs Szathmáry galvanized biologists to seek principles in such transitions, a significant number of which involve an increase in the degree of nestedness.[4] Bronowski had earlier noted one such transition: cells evolved to animals composed of cells.

An aspect of Bronowski's insight was that the process of combining and integrating could link the fields of physics, chemistry, and biology. His friend and colleague Jonas Salk, of polio vaccine fame, even extended the concept of building up into culture.[5] This is also my purpose here. Though the older idea from cultural scholars that humanity progressed in a distinct series—small bands enlarged into tribes, and tribes grew into states—has been judged too simplistic, we can acknowledge that over human history social scale remarkably increased and created more complex nestings.

In *Consilience: The Unity of Knowledge*, the biologist and writer Edward O. Wilson developed the metaphor of a thread that could be passed seamlessly from physics to culture.[6] I won't get into any hurried debate about reductionism versus holism here (good science—and any good scholarship, I would add—works with both ideas), but one of Wilson's concerns was to point out where the thread of our understanding has gaps as worthy sites for young researchers to position themselves in.

If we combine Wilson's thread of connectivity with Bronowski's stratified stability and the ongoing, vigorous wrestling with concepts of nestedness in physics, biology, and cultural studies, we might expect a worthwhile quest would be to examine transitions and levels not just within any one large field but also across the whole shebang of phenomena from physics through biology to culture. I made such an examination for this book, and I call this run of transitions and levels the *grand sequence*. The grand sequence crosses the foundational structures of physics, the wonders of major inventions of life and evolution, and the majesty and enigmas of major cultural transitions.

Here I also offer a dose of intellectual thanks to systems theorists and thinkers—for example, those who have forged new territory in the topics of complexity, emergence, self-organization, networks, and all related inquiries that seek descriptive or formal mathematical commonalities across nominally disparate fields or phenomena.[7] I greatly admire this kind of work. But I do not engage in parsing it and comparing it point by point to what I am proposing here. For example, I do not regularly use the terms *emergence* and *synergy*, which are used elsewhere.[8] My focus is not on general emergence, networks, or synergy as a topic or topics (vast!).

I am instead trying to do something rather simple, almost architectural. Can I find a foundation? If so, what does it have that allows the walls? Then what do the walls have that allow the roof? I want to ask about *things and relations* in a grand sequence from quarks to culture, using combogenesis as an iterative or rhythmic theme. Can we define fundamental levels? You might think of these levels as specific categories of emergence or perhaps as the major transitions of biology extended inward into chemistry and physics and outward to culture.

I have done a share of math modeling in my work with the global carbon cycle and advanced life support,[9] but I'm not doing math here. It wasn't relevant for my architectural approach. I instead try to stay on course with the logic of the concepts I propose and discuss: innovations of things and relations at each level. I naturally try to draw on the best scholarship wherever I think I have found it. If there is anything new here, it's in my synthesis. And perhaps it's in the webs of questions the approach leads to. It has been a thrill for me to attempt to balance my integrative synthesis with an examination of the state of knowledge in various fields.

The book has three parts. Part 1 introduces the goal of defining an overall sequence from quarks to culture and develops the general concept of combogenesis, the process of combination and integration of things that results in levels, each with new things or systems that possess new relations.

Part 2 unfolds each of twelve levels derived from combogenesis. I provide my understanding of the cosmically patentable accomplishments achieved at each level. How did these accomplishments facilitate the next event of combination and integration? I aim to weave established science with more than a few outstanding mysteries. Just attempting to review the knowns and unknowns of these levels made me revel in every scale of nature and culture so much more than I had previously. I tend to be a reveler in these matters, anyway, so it takes a great deal to ramp up my enthusiasm even higher. I hope some of this feeling gets conveyed to you.

Finally, after I have fleshed out the levels in part 2, I seek out additional themes and parallels across those levels in part 3. I propose a trio of *dynamical realms* in the grand sequence: the realm of physical law, the realm of biological evolution, and the realm of cultural evolution. Each realm consists of multiple levels and its own special base level. I use these realms within the grand sequence to reinterpret generalized evolutionary dynamics, aided by a pattern that I call the *alphakit* because it is both genetic and linguistic. And there is more to uncover by using the special base levels as anchors for novel inquiry. Finally, I ask if my findings can help frame what is happening in today's world, where a transformative future seems to be manifesting itself at an ever more rapid rate.

I hope my readers are in the creative class in the widest possible sense, from artists to complexity theorists, intellectuals in all fields, "big-history" educators, social scientists, humanists, natural scientists, systems thinkers, technologists, writers, musicians, lay scholars, spiritual seekers—anyone with passionate interests in the big picture. This is easy to say, but

I mean it. For instance, I have always warmed to opportunities at my university to teach courses that bring me in contact with students from a diversity of fields and interests. I also hope that those interested in popular or semipopular writings in physics, biological evolution, and human origins will want to join in my exploration of the grand sequence and its implications. I hope to avoid any grievous injustice to anyone's specialty. I know I have left out favorite nuances and debates, but I hope any slip-ups will not affect the general argument.

I have tried to write in a welcoming, jargon-free style, carefully defining and deploying any key new concepts. I use technical terms for findings crucial to the relevant fields of scholarship. I hope I have adequately guided readers through the main concepts of those fields by sticking to results. Also, I had to coin some terminology, but only when I knew I would be using a given new word often and would gain from having a shorthand for certain key concepts developed in the book.

I have written this book in part to satisfy a dual longing: for a narration of how we came to be and of our place in the lived universe as well as for the logic and operating principles that have powered that narration. In this work, these operating principles are very general. Perhaps they will seem philosophical. Fine! I sometimes think of them as topological-functional, based in words and images that describe patterns of systems. However one might categorize them, I use the principles to derive and explore a sequence of levels from physics to life and biological evolution and on to our legacy of cultural transitions. May any answers in this work also serve as guides to more questions, with the potential for opening awe and individual inquiry.

QUARKS
TO CULTURE

1

COMBOGENESIS AND A GRAND SEQUENCE

——

How did the simplest things in the universe transform into the riches of culture we have today? Can a natural narrative provide an answer? As a metaphor, consider things that build one after another by events of a special class. At each event, smaller things (systems, entities) from prior levels combine into larger things on subsequent levels. This general process is here called *combogenesis*: the births of new types of entities by the coming together and integration of prior things. Taken altogether, the fundamental or basic levels of combogenesis are called the *grand sequence*.

1

NATURAL CHAPTERS
AND NESTED SCALES

SUMMARY: Within the body are living cells, and within the cells are atoms. Going inward in scale of size takes us back in time to the first origins of these fundamental types of things. Can this simple insight provide us with a general logic that we might use to derive natural chapters in a narrative of the universe?

A UNIFIED NARRATIVE?

Who are we? Where did we come from?

As a start, consider the flaring forth of space, time, energy, and matter at the Big Bang. Billions of years later Earth's pageantry of life thrived and evolved, from ancient wriggling microbes to today's roaring elephants. Just a hair's breadth in time prior to the present, the florescence of conscious human minds eventually created today's rich tapestries of culture. How did all this happen?

Physicists, biologists, anthropologists, and others with viewpoints based in specialized fields can offer answers for portions of this great saga. Is there a way to unite all those parts into an integrated narrative? Even more radically—and at the heart of this book's findings—can this narrative contain what might be considered natural chapters? Is there a way we can ask—poetically of course—the universe to be our narrator?

NARRATIVES WITH RHYTHMIC THEMES

Narratives that progress by the beats of general internal themes run through many of the world's ancient creation myths. The Bible has its famous seven days of creation. The Aztecs envisioned a series of worlds or "suns" formed and then destroyed. Native Americans of the Southwest tell about a final emergence to us, upward through a *sipapu*, or portal, perhaps inside the Grand Canyon, that followed previous climbs upward with help from animals.

What these myths share is a succession of stages—in the form of days, suns, portals—structured by a theme that operates in a repetitive cycle.[1]

In contrast, a science-based narrative of where we come from seems to lack any simple rhythm, but there are plenty of main events. An astrophysicist would hail the point when the first atoms condensed not long after the Big Bang. A paleontologist would get us to gasp at the colossal impact from space, which unleashed energy equivalent to a billion Hiroshima-leveling atomic bombs, extinguishing the dinosaurs and 90 percent of all species from giant to microscopic. An archaeologist might regale us with the epic of Gilgamesh from ancient Mesopotamia, the earliest recorded awareness of personal mortality.

Experts will have no problem identifying major events. But how are those events determined to be the truly main ones? It seems to depend on the narrator and field of expertise.

There is a new way to make this determination, I suggest, one that has the potential for revealing an overall, integrating narrative with a rhythmic theme that defines main events. Here the goal is not to invent a theme. But does a theme present itself? If so, can it help us to understand our place in nature?

The theme to be proposed, unlike the narrative devices of the ancient creation myths, must tie together hard-won scientific knowns. As it turns out, we will find a succession of main events like those in many of these myths. The events did not arrive at equal intervals of time like months in a calendar or pulses of a song. Yet we will see a rhythm of sorts—let us say a highly irregular cosmic heartbeat. And the results of each beat were more cumulatively creative than the beats of your heart. Each beat produced a fundamental new level of being. We contain some of those levels. And we live within others.

LOOKING INWARD TO YOUR BODY'S LEVELS OF THINGS

I start with a thing immanently real, your own body.

Your body is a major type of thing. Billions of other individuals of your same type—*Homo sapiens*—currently walk Earth. Indeed, you are a member of many sets that increase in generality: You are an individual human body. You are an individual animal body. You are an individual multicellular organism. You are an individual life form. And at a scale of generality shared even with quarks and atoms, you are an individual thing.

Inside your body live microscopic things called cells, more than 30 trillion of them.[2] (Let us not count the bacteria living on and inside you.) If we ignore the issue of mutations, your cells carry the same DNA, yet they radically differ in size, shape, and function across your body's tissues and organs. During their choreographed dance that began with a single fertilized human egg cell, the exploding populations of cells communally affected each other. The result guided certain genes and thus proteins into abundant expression in some cells but not in others. Thus, you are constituted as a vast community of muscle cells, nerve cells, bone cells, liver cells, and many other types of cells.

Let us travel more deeply down into the microcosm. What's inside cells? Both DNA and proteins are examples of the type of things called molecules. Your cells typically contain tens of thousands of subtypes of proteins. Furthermore, like all molecules, those proteins contain atoms. So we can logically say that because molecules are inside cells, and because atoms are inside molecules, it's the case that atoms are inside cells. Because a typical cell of the human body contains about 300 trillion atoms,[3] then (very roughly, within a factor of 10) there are about as many atoms in each cell as there are cells in your body.

Physicists and chemists recognize atoms as an extraordinary, basic type of thing. The properties and behaviors of atoms are relatively well studied. At this point, to keep it simple, I focus on atoms as a crucial part of the logic but come back to molecules in the next chapter when I need to better refine the search for and definition of basic levels.

As noted, there is something fundamental about each of these scales of things: (1) human body, (2) cell, (3) atom. There is no debate here about the essential roles each plays in our biological lives. Each is also essential to our understanding of who we are. Furthermore, each type is quite general, for each has many subclasses: many different individual humans who have many kinds of cells that contain many kinds of atoms. We might lump the members of all these types and subclasses into one gigantic, inclusive set of "all things," but then we would have no distinctions to work with. So I propose that in the search for natural chapters in a narrative of the universe, we look for some of the most general types of things that are vital to our existence—for example, the types that form the basis for some of the earliest lessons typically learned in science classes. Atoms and cells surely headline that bill.

SYSTEMS AS THINGS: TERMINOLOGY

There is no perfect term for what I want to discuss. So far I have been talking about types of "things." And I have shown that these things tend to physically contain smaller things inside them. A thing made of smaller things, or components, is often called a "system." The point made when using this word *system* is to emphasize that bodies are systems of cells, cells are systems of atoms, and atoms are systems of nucleons, and so on—in any case, all contain coordinated, interacting internal parts. We have not yet gotten to the guts of the atoms, but we will when we have more purpose.

A word about this word *system*. It's a good alternative to *thing*. So I do use it. But when it is overused, it can feel a bit cold. It invokes mechanism. It emphasizes the fact of a thing having parts but doesn't embrace, well, the whole thing. But does the word *thing* feel any warmer? How about *entity*? I have used the latter and will continue to do so. But it's not the best term. It conjures spooks. One might try to coin a new word. I was going to shorten *entity* into *ent* until it was pointed out to me that Ents are a race of creatures in the *Lord of the Rings* trilogy. We might tinker with *ontology*, a term with ancient Greek roots used for the general study of things and existence. So *ontum* perhaps? Then, we could state that an ontum (say, a cell) is made of smaller ontums (say, atoms).

The open-ended, nonspecific, plain-wrapper blandness of the word *thing* might give it some advantage here in the quest for a preferred term.

But I employ *thing, system, entity,* and occasionally even *ontum* as synonyms. Whatever the word, what is important for the logic is the fact of multiple scales of what can be called *nesting* or *nestedness* in the most fundamental levels of being: systems nested inside systems, things nested inside things.

Note that in the examples given, the existence of each scale of smaller thing is necessary for the existence of the larger scales of nestedness. The body does not just happen to have cells; the body is alive *because* it is a coordinated system of living cells. To exist, our cells need the deeper-down building blocks of atoms.

GOING BACK IN TIME BY GOING SMALLER IN SCALE

To more richly appreciate this nesting of fundamental ontums, let us consider *time*. As we go inward—or downward (feel free to choose your own term for this particular form of movement)—the deeper, smaller scales of the microcosm take us back in time. Going inward is a form of time travel. How so?

For focus, we can ask about the origin of any particular type of thing. Consider multicellular animals, of which we are a member. Current estimates from the science of biological evolution place the origin of multicellular animals somewhere between 700 million to a billion years ago.

Next down in the scales discussed so far are the animal cells. Consider them from the perspective of origins. They, like the cells of all animals and of plants and fungi, too, are members of a certain, particularly crucial and complex type of cell that evolved roughly 2 billion years ago, perhaps a bit earlier (much more detail on that membership is given later in the book). Though the birth date for this wonderfully successful cell type is still debated, it certainly came before the evolution of animals. We know that because this same, general, complex type of cell is found in all animals, plants, and fungi. It is therefore ancestral to all three (and more) forms of multicellularity and thus to us, *Homo sapiens*.

How about the incomprehensibly tiny atoms? As noted, an astrophysicist would hail their first origin as a major event in the narrative of how we came to be. The first atoms came into existence when the average temperature of the expanding universe cooled to about that of our sun's surface, roughly at the 400,000-year mark after the Big Bang, around 13.8 billion years ago. And the specific atoms of your body, at least most of the atoms in you right now, originated during ejections of matter from supernova stars from eras prior to the formation of Earth 4.6 billion years ago.

Therefore, as shown in figure 1.1, a logical sequence of nestedness correlates with first origins of each type or scale of each thing in time. A sequence of creation in time is also a sequence of enlargement in the sizes of the systems created, illustrated by nested scales of bubbles. And all of these levels coexist right now in your living body.

How each type was born is a distinct story, of course. Those fascinating details will matter a great deal later in this book. But not yet. Certain variations among subclasses or subtypes are important, too. For now, though, the point is simply this: for each general type of thing, there was a time when it did not exist, and then it originated; after that, it did exist and has ever since.

Perhaps think of the narrative like this. At one time early in the history of the universe, the atom was the most newsworthy, avant-garde thing around. Then billions of years later, the complex living cell emerged,

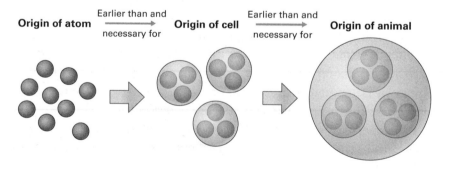

FIGURE 1.1

The origins of three fundamental types of things creates a sequence of an expanding nesting of systems. Each circle or bubble simply symbolizes a generic type of thing or system, regardless of scale.

at least on Earth, as the hottest avant-garde system—not only the newest type of ontum but also the ancestor of all cells of our bodies as well as of all cells of mushrooms and trees. And those ancestral cells were able to come into existence only because atoms already existed—the atoms could take on roles as atomic letters within the intertwined chemical reactions inside the cells. Finally (so far in our logic), at some point after the origin of complex cells, the multicellular organism originated as the avant-garde entity, the most newsworthy thing on Earth, with a special nod to the branch we call animals. A sequence in time was built by nesting in series from smaller into larger.

■ ■ ■

Can there be a science or a study of everything? When physicists talk about their quest for a theory of everything, they do not include the works of Shakespeare. To physicists, a theory of everything would unify the understanding of fundamental subnano building blocks and connectors of matter, energy, and especially the different basic forces. But what if we *were* to desire to include Shakespeare along with the quarks of physics as a member of the set of things being studied? What if we were to truly embrace *everything* in a study of everything? Then, would the phenomena being studied involve *pattern* itself? I think so. Everything that can be studied has pattern, from atoms to societies.

In this book, I am focused on innovations in the patterns that built up into the pattern called the human being, but, as will become clear, I am interested not just in the human but in humanity. This buildup might seem important only from our viewpoint. But there are other vital systems that are also very ancient, such as the mighty Earth as a planet. Such systems preceded the human body. "Don't forget Earth!" my geochemist colleague Peter Westbroek told me. Or we might pay respect to the entire cosmos itself, which preceded any atom. Since the Big Bang, the universe's all-encompassing stature has never diminished. In fact, size-wise, the universe continues to grow with the cosmic expansion, which is known to be accelerating. Yes, we are nothing without the cosmos. And we would be nothing without Earth. Every day we breathe Earth's air and drink its waters. So what about these huge systems that preceded us?

Although the cosmos and Earth were vital to our creation (and remain vital to our existence), they were not things on the direct line to us, in the sense of classes of systems that progressed in a sequence from small to large that took atoms to cells and cells to animals. Put another way, cosmos and Earth are not entire whole things that are literally internal parts of us. The atoms of the cosmos—but not the cosmos itself—are inside us. The molecules of Earth—but not Earth itself—are inside us. Atoms and molecules, as types of very much smaller things within us, *are* on the direct line of the narrative we will focus on here.

Why confine the search in this way? The goal is to find the basic chapters in the narrative of who we are and how we got here. The clue to discovering these chapters might be in this building up from small to large in a nested sequence of systems or things. It is only a clue, however. We would still need to refine a logical model that can be engaged to find the sequence we seek.

2

THE CORE THEME

—

Combogenesis

SUMMARY: A concept is offered: combogenesis. *Events share a certain kind of rhythm in the creation narrative from quarks to culture. Each event combines and integrates prior things into new, larger things. Each forges a new level of type of thing (system, entity, ontum). Overall, as a series, these fundamental levels constitute a grand sequence. New relations possessed by new systems at a given level establish possibilities for systems on each next, subsequent level. The aim is to use combogenesis as a framework to proceed deliberately, to synthesize across various fields of knowledge, and to explore how fundamental types of things and relations came about in time.*

SEEKING FUNDAMENTAL TYPES

Consider molecules: as noted, clearly one of the major types of things. For the origin of life, biologists posit that cells originated not directly from combinations of atoms but rather from combinations of molecules. Those molecules contained atoms. So, refining the logic, to the sequence of atoms, cells, and animals shown in the prior chapter, the origin of molecules should be added in between the origins of atoms and cells. This is done in figure 2.1.

The figure shows the nested bubbles again. That style of diagramming levels would soon get overly complex as levels are added. So the figure also shows an alternative method I often use: concentric circles. Individual

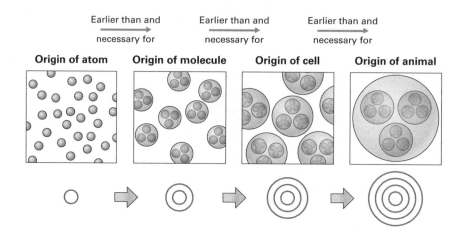

FIGURE 2.1

Because atoms had to exist before they could connect into molecules, and because molecules had to exist before cells, we can place the origin of molecules as a new stepping-stone between the origins of atoms and cells.

layers of circles from small to large represent small things (as types) nested inside larger things (as types).

What do we learn from this insertion of the molecule's origin into the sequence? Will there simply be an infinite number of possible insertions? I don't think that is the case. But we do learn that we require a more explicit method or model to guide how to determine the main events in which types of entities came into existence.

COMBOGENESIS

Wouldn't it be cool to be able to know, roughly, the main events of "coming together" that took place from the Big Bang to us? Such knowledge could help inform the narrative we tell by way of chapters that yielded the universe we live in. Such knowledge, in other words, could be central to an understanding of how we came to be.

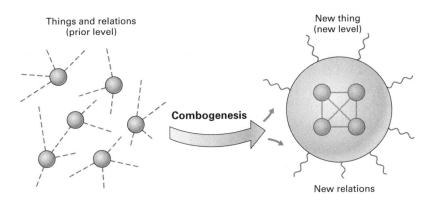

FIGURE 2.2

The circles on the left are the things (systems, entities) of a prior level. The dashed lines represent the relations they have with each other and with all things in their environment. The single larger circle on the right is the new type of thing (system, entity) on the new level, which results from combination and integration, *combogenesis*; the wavy lines radiating from the circle represent the new kinds of relations it has and is capable of having.

A phrase such as "coming together" is a bit awkward, despite a certain sex appeal. And I don't want to be limited to even wordier phrases, such as "smaller whole systems merging into new, larger whole systems." A single word for the theme should help keep the flow of logic on track when the investigations start digging into variations of detail in the types of things.

So I use the following word: *combogenesis*. The idea is genesis by combination. The word riffs off the informal term for a small, cooperative jazz group, *combo*. Why coin a term? Quite simply, I haven't found any current word for the concept I want to discuss.[1]

Combogenesis is the genesis of new types of things by combination and integration of previously existing things. See figure 2.2, which shows the basic features to be described.

In this work, the word *combogenesis* is applied *exclusively to the major new types of entities* generated along the sequence that includes the most fundamental types of things from quarks to culture. You might decide to use the word when you cook and combine ingredients in a recipe. But if I have coined a term, I can restrict its use at least in this book. And as I have said, though Earth and the cosmos are vital, I'm not interested in all systems, only the special ones that you will see as you read.

To flesh out what intrigues me requires a few additional terms. Again, see figure 2.2. I have already been using one term: *level*. It appears, for better or worse, more than five hundred times. I needed a word to designate a type, a category, a set of fundamental things or systems at various scales of sizes. Examples used so far are the level of the atoms, the level of molecules, the level of cells, and the level of multicellular organisms (which include animals, plants, fungi). An event of combogenesis takes participants from a previous level and integrates them into a new general type of thing on a given new level.

A certain distinction is important: once new things reach a certain level, additional new kinds (subclasses) of those new things can be and most usually are created within that same level. Thus, in this book the evolution of multicellular animals from cells *is* an event of combogenesis, but the evolution of mammals from reptiles *is not* because that change took place *within* the level of multicellularity. As another example, consider subclasses within the level called molecules. Although molecules can be made directly from atoms (members of the previous level, which are at a smaller fundamental scale), *most* molecules within the vast complex landscape of their subclasses get forged during chemical reactions among other subclasses of molecules and thus all within the level of molecules.

NEW RELATIONS

With these concepts of events of combogenesis, levels, and new types of things, the overall scheme of combogenesis is utterly different from, say, a gravitational buildup. Consider large planets forming from gradual accretion of smaller gravity-possessing masses, a process that begins with tiny dust particles in the formation of solar systems. In this progression, during the buildup, gravity merely becomes greater and greater in overall magnitude. (Yes, it's a different story when a star ignites.)

In contrast to a gradual increase in gravity during a planet's accretion, in the phenomena I seek to bring to light by applying the model of combogenesis, the new things at each level are much more than just larger in size. They are *new* phenomena. Of course, what is new and what is meant by new at each level will have to be made clear as we progress. But one portion of our inquiry that should never be neglected is the characterization of the *new relations* that come about with the new things on each new level.

Things have relations with other things in their surroundings. By "relations," I mean whatever interactions a thing has or is capable of having with other things.[2] Relations include, for example, the forces of physics, a cell's importation of nutrients, a rabbit's ability to run away from a coyote. Relations include people talking or taking actions first vetted in their imaginations. Relations differ among things, and to a large degree they brand those things. As far as we know, atoms do not talk (in language), and though the force of gravity exists among people because they have masses, it can be set aside for, say, most discourse of politics. To a large degree, the narrative of quarks to culture is an awesome development of innovations in relations.

In figure 2.2, dashed lines represent relations possessed by the smaller things on the left. Wavy lines branching out from the "new thing" on the right, after combogenesis, show the relations it possesses. The contrast between the two styles of radiating lines, dashed and wavy, symbolizes the creation of new, unique relations by the things at each new level. Again, those wavy lines on the right will prove key to the analysis (and to what happened in our universe), and I will have much more to say about the new relations in part 2 once the evidence is in about what they are and how they operate for the levels examined.

One reason the production of new relations is so important is that this feature of the concept of combogenesis leads to a way to have iterations or rhythms of the general process of combination and integration. The new things with their new relations on the right side of figure 2.2 will become the things and relations on a new (imagined) left-hand side in the next iteration at a subsequent and even larger scale. The process can be considered helical: a sequence of change or development of new things and relations is generated through the repeating, cyclic theme.

THE GRAND SEQUENCE

I introduce one more special term. As noted, the result of this process, applied recursively or iteratively—to say it once more in slightly different words—creates out of the integration of prior, smaller-scale systems a nesting of increasingly larger systems. The result: a sequence in time is linked to a certain kind of nestedness in space. (Recall that going inside the body is going back in time.) Now I suggest that this sequence itself deserves a name. At the very least, I need to be able to refer to it.

The sequence could be considered a path of the construction and creation of things from the simplest particles of physics to us. But a philosopher-colleague cautioned me about the term *path*. It implies, said Bill Ruddick, that the universe's creation of things followed some preestablished, perhaps preordained path. This is not true as a judgment I would be willing to defend. Yet neither is it without at least a smidgen of validity: to an extent, certain beats of the theme were more inevitable or causally necessary than others, as I hope I can make clear.

Trying to be sensitive to terminology, therefore, for the entire sequence itself, I offer a word borrowed from a wonderful book by the physicists Stephen Hawking and Leonard Mlodinow titled *The Grand Design*.[3] I call the sequence, with its overall series of levels, the *grand sequence*. Those two words do not have to imply a predetermined path. It is a sequence. And it is grand. It got to us.

SYSTEMS LARGER THAN THE HUMAN BODY IN THE GRAND SEQUENCE

How many basic levels of things were created in the grand sequence by the rhythms of combogenesis? What was new about each of these levels? What innovative relations did the new things have that made it possible for each subsequent event of combining and integrating? Suggested answers to these questions require applying the theme, which I do in subsequent chapters.

But, first, a note on something implied in this book's title. I have mentioned the term *culture* several times yet have stopped the analysis in the figures so far at the animal body, of which we are examples. Let's move outward. Considering our bodies as constituents within other, larger systems surely brings us to cultural systems or social systems or what some might prefer to call "sociocultural systems." In this book, I simply use the word *culture*. I suggest several levels of culture in the grand sequence.

Individual people nested within a much larger group is strikingly illustrated in the famous frontispiece to Thomas Hobbes's seminal book on political theory, *Leviathan or The Matter, Forme, and Power of a Common Wealth Ecclesiasticall and Civil* (1651).[4] The image in figure 2.3 shows the geopolitical state in the symbolic form of a giant, crowned superbeing. His body contains the people. The idea is that the people are parts of a national or political body. The relationship between the people and system, embodied in the image of the ruler,[5] is the subject of Hobbes's book,

FIGURE 2.3

Portion of the frontispiece of *Leviathan* (1651) by Thomas Hobbes, with one part magnified to better highlight the people, who are like cells in a body. Here they are inside the geopolitical state, symbolized by the giant ruler responsible for holding the system together, for better or worse.

and this relationship is still a heated heart of political debates all over the planet. But for our purposes here, we simply witness this picture as social nesting. If you are roughly typical, you are a part inside larger things, whether you like it or not (depending how your work day is going)— Freud's *Civilization and Its Discontents*.

Humans have been physically human for roughly 200,000 years. Thus, the human body preceded the creation of larger social units, such as the sprawling geopolitical states of ancient Sumer, Peru, and China, all of which were more recent than 10,000 years ago. Are we to the giant systems we live in like cells to bodies? Or like atoms to molecules? Certainly in each nested pair, the smaller originated before the larger and is contained in the larger (atoms in molecules, cells in bodies, human bodies in geopolitical states). But did the grand sequence go from the human body directly to the geopolitical state? Or were other levels of fundamental things created in between? Will we find that we need to insert other levels, similar to the way we inserted molecules between atoms and biological cells?

■ ■ ■

The plan is to use this model or theme of combogenesis to investigate fundamental types of things as chapters of creation. Please note this is a logical model, not a mathematical one. And although I developed it to let the universe be the narrator, it is not possible, of course, to avoid my personal subjectivity in applying the organizing theme. My most questionable choice, I believe, was to start this quest itself—that is, to seek a theme that can synthesize a unified narrative from quarks to culture based on well-established facts. Of course, and apropos to caveats and confidence about my suggestions in this book, I am open to improved precision in the future, from other thinkers and as new findings from the disciplinary fields emerge, to help distinguish types of things and possible intervening levels among the things.

This scheme might seem very anthropocentric—this defining of a "grand" sequence as the concentric rings that led from simple things of physics to *Homo sapiens* in their modalities of cultural being. It *is* anthropocentric. Heck, yes, sure it is. A central target of ancient creation myths was to explain how humans came to exist here on the cosmic scene. And today we, the living, are no less curious to find an answer to this grand question.

To satisfy this curiosity, I hope it is clear that I intend to use the concept of combogenesis as a tool to help address issues that arise in attempting to discern the main transitions and levels of the grand sequence. That will take place in part 2. Levels and new things with new relations—all will be the subjects of an inquiry whose range is both inside us and outside us.

In part 3, assuming part 2 provides us with a superfamily of phenomena that share the property of being levels in the grand sequence, I start asking about connections among the members of that superfamily. In other words, we might eventually seek connections among subthemes within the theme of combogenesis. Perhaps we will understand a bit better the overall pattern-based, theory-of-everything nature of this amazing reality we inhabit and have helped form.

2

TWELVE FUNDAMENTAL LEVELS

Postulating a rhythm in a natural narrative from quarks to culture leads to questions: Can specific levels of the grand sequence be identified? What things with relations came "up" from the previous level as potential players in each next step of combination and integration? On each level, what were the most crucial new relations possessed by the new things? How exactly were those relations capable of making the new things combine into still larger things of the next successive level?

Here, in part 2, I make cases for twelve levels in the grand sequence and offer answers to these questions for each of these levels. Each level gets an individual chapter so that I can focus on the most relevant aspects of innovations of things and relations. I aim for consistent descriptions, using the concepts within the overall theme of combogenesis and lingering a bit on aspects of each level that I have found fascinating as well as special. As a result, I hope to show that the logic of combogenesis does warrant the concept of a grand sequence made from a repeating theme.

After that, in part 3, I focus on a second aim. I turn to other themes shared by some but not all levels within the grand sequence. But we can't examine those themes until after we have more "data" on the twelve levels here:

1. Fundamental quanta
2. Nucleons: protons and neutrons
3. Atomic nuclei
4. Atoms
5. Molecules
6. Prokaryotic cells
7. Eukaryotic cells
8. Complex multicellular organisms
9. Animal social groups
10. Tribal metagroups
11. Agrovillages
12. Geopolitical states

3

A BIG BANG START OF
THINGS AND RELATIONS

SUMMARY: This deep base level to the grand sequence consists of the set of fundamental quanta that make up what physicists call the Standard Model. The model has an ordered array of quarks, electrons, gluons, photons, and other basic entities. Built into the fabric of our universe, the quanta in the array are organized by a primordial twoness: matter-field quanta and force-field quanta. This remarkable start with basic "things" that interact is beaming with prospects for combogenesis.

THE ENIGMATIC, FUNDAMENTAL QUANTA

In the beginning, as currently known by astrophysicists, the Big Bang's primordial burst came about 13.8 billion years ago. As the energy density of the universe started rapidly to diminish, like orders arriving from a fixed menu, the fundamental quanta emerged into existence in rather quick succession. These basic things were unimaginable in count but relatively few in terms of their basic types.

Modern physicists describe these quanta with an enormously successful theory called the Standard Model. Frank Wilczek, physics Nobelist, says that "Standard Model" is a "grotesquely modest name for one of humankind's greatest achievements."[1] Given such awe, I am tempted to call this theory the Magical Mojo Model. But here, in honor of the legacy of terms accepted by experts, the Standard Model it will be.

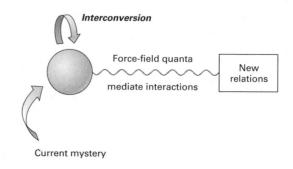

Matter-field quanta: quarks, electrons, etc.; and force-field quanta: gluons, electrons, etc.

Interconversion

Force-field quanta
mediate interactions

New relations

Current mystery

FIGURE 3.1

The starter level to the grand sequence: the Standard Model of particle physics, with two sets of fundamental quanta: matter-field quanta force-field quanta ("things," but they are also "relations"). A deeper, prior level is suspected but not established (the "current mystery"). The term *interconversion* is used in these figures as a shorthand for processes that change types of things within the new level, as discussed in the text for each specific level. Here interconversion is specified according to the laws of particle physics.

What are these quanta? Even what to call them can be contended. Often designated the fundamental "particles," they are not particles in the usual sense we nonphysicists would likely understand. Any of the quanta can be real or virtual. In their virtual manifestations, they flash in and out of existence, from the vacuum or from interactions with real particles or even with other virtual particles. Some quanta have mass; some are massless. None of them has a known size. I adopt terms from physicist Bruce Schumm,[2] who calls the gamut of these enigmatic things "quanta."

In addition, within the Standard Model the quanta are divided into a primal and complementary duality—technically, the fermions and the gauge bosons. Again, in Schumm's evocative phrases, the two types are (1) matter-field quanta and (2) force-field quanta. This division into two basic subsets within the overall set rules over how the quanta enable the first event of combogenesis. Therefore, after giving some crucial and provocative background, I get into how they work together.

Here I list the set of the fundamental matter-field quanta and force-field quanta in these two categories. For our purposes, not all members are equally important. But some, as explained, stand out as main players. And all are vital as members of the Standard Model, an integrated, empirically verified theory. Some names are recognizable even to nonphysicists. The Higgs boson, an integral part of the Standard Model, does not fit in either of the two categories, but it is shown in the more detailed Standard Model in the first figure of this book's center insert. Note that the matter-field quanta come in groups of three, which physicists call "generations."

UP-QUARK-LIKE MATTER-FIELD QUANTA, IN INCREASING MASS FROM LEFT TO RIGHT

up quark	charm quark	top quark

DOWN-QUARK-LIKE MATTER-FIELD QUANTA, IN INCREASING MASS FROM LEFT TO RIGHT

down quark	strange quark	bottom quark

ELECTRON-LIKE MATTER-FIELD QUANTA, IN INCREASING MASS FROM LEFT TO RIGHT

electron	muon	tau

NEUTRINO-LIKE MATTER-FIELD QUANTA, ORDERED WITH ELECTRON-TYPE PARTNERS

electron neutrino	muon neutrino	tau neutrino

FORCE-FIELD QUANTA

gluon	photon	W and Z gauge bosons

Note: The Higgs boson, though now in the Standard Model, is not in this set.

A couple comments of particular relevance:

The quantum world is not like the situation with human height, weight, or eye color, which can be infinitely varied. All electrons have exactly the same magnitude of electric charge. All up quarks have exactly the same magnitude of electric charge (different from the electron) and color charge (more on color soon, which electrons do not possess). Indeed, we define the types of quanta by the invariance of properties within each type. The quanta seem reduced to a minimal complexity, with just a few precise mathematical qualities.

Down in this abyss of the nano–nano, the "things" might be considered mathematical objects. Werner Heisenberg (1901–1976), reflecting back on his discoveries of quantum mechanics, said he "felt almost giddy at the thought that I now had to probe this wealth of mathematical structures nature had so generously spread out before me."[3] Living physicist Max Tegmark emphasizes that the basis of reality *is* mathematics.[4] If only the English poet John Keats were still with us. In "Ode to a Grecian Urn," he memorializes the aphorism "Beauty is truth, truth beauty." Today, Keats might be moved to write "Ode to the Standard Model," concluding that "matter is math, math matter."

The fundamental quanta are a weird opening act to the grand sequence. Of the quanta that have mass, physicist Lisa Randall calls that mass "an enormous mystery."[5] For physicist Martinus Veltman, the "greatest puzzle of elementary particle physics today" is the three generations of each of the types of the matter-field quanta.[6] There are also ongoing questions about fundamental constants whose values come from measurement but are not understood by theory.

Physics Nobelist Richard Feynman famously said when describing the universe implied by the weird findings of particle physics, "If you don't like it, go somewhere else—perhaps to another universe where the rules are simpler."[7] No worries, Richard, we like it!

QUARKS IN US

We turn now to the general type of matter-field quanta called "quarks" for a bit of detail because quarks are key players to watch in the next level.

Like all the quanta, quarks exist in a realm bizarre to the typical daily lives of coffee, cars, and cash. Nevertheless, perhaps because of electricity, few people have problems with the concept of one of the other types of matter-field quanta, the electrons. Even fewer take issue with the concepts of the much larger atoms or molecules. But the mention of "quarks" often gets puzzled looks. Yet without quarks there would be no atoms, no molecules, no living cells, and definitely no coffee-stoked, cash-carrying, car-driving people.

As you can see in the tables outlining the types of quanta and their names, quarks have been bestowed with whimsical names: "up," "down," "charm," "strange," "top," and "bottom." The physicist Murray Gell-Mann borrowed the fabricated word *quark* from Irish writer James Joyce. Quarks come in two subtypes (up and down), each with three generations. And they all have their antimatter counterparts, the antiquarks. Furthermore, each quark can possess at any moment one of a trio of different color charges (red, green, or blue) because a quark alternates colors during interactions with other quarks and with gluons. I say more later on this color that is not real color.

Their sizes? Not established. Experiments have been able to determine that if quarks are sizable, they are smaller than 10^{-18} meters (a billionth of a billionth of a meter).[8] Quarks might be "points." What does that mean? A scale so infinitesimal is a pain to wrap your head around. But try shrinking your head, first to the size of the period at the end of this sentence. And then make that same relative jump in scale about six more times to reach the 10^{-18} meters scale. That's the maximum. The reality is likely even crazier.

All quarks have mass. But why they have the precise masses established is an ongoing enigma for physicists, as Lisa Randall emphasizes.[9] Their masses do come into the story in the next chapter and in a way that I think you'll find mind blowing.

In addition to mass, all quarks possess electric charge. That characteristic seems at least a bit familiar. The electric charge of an electron is defined as −1 unit (the "negative" is only a convention to contrast it with the complementary property of a positive electric charge). Relative to an electron's charge, quarks have an either +⅔ or −⅓ electric charge. So quarks' fractional electric charges are odd. The enigma there is fortunate, though, for, as becomes apparent over the next few chapters, what happens when quarks combine works out nicely for us humans.

All quarks possess another basic attribute called "weak isospin." Like electric charge, weak isospin is a conserved property during interactions. It has the same mathematical behavior as angular momentum, but physicists emphasize that we should not think of quarks as literally pirouetting. Weak isospin facilitates the conversion of some types of quarks into other types during certain interactions. Fortunately for us, though weak isospin is crucial to the universe, it is not crucial to the main focus of our narrative.

Finally, of huge consequence, quarks possess color charge or, simply, color. The scene gets increasingly bizarre. The term *color charge* does not at all refer to our familiar visual color but rather is physicists' playful appellation for a basic property possessed by all quarks in equal amounts. In the well-vetted theory of quantum chromodynamics, there are three "flavors" of color: red, green, and blue.

In our familiar world, we have experience with electric charges. Sparks fly if we touch two ends of a wire connected to the two poles of a battery— good to recall if we are lost in the woods, need to make fire, and have a battery and proxy for a wire. Electric (or electromagnetic) charges come in pairs: two poles to a magnet; two ends of a battery; electricity with its positive and negative charges; positive sodium and negative chloride ions from table salt dissolved in water (indeed, many substances dissolve in water specifically because they separate into negatively and positively charged ions).

In contrast, the tripleness of quarks' color charge might seem very odd to us because of our familiarity with this doubleness of electromagnetic phenomena. And unlike an electron that has only negative electrical charge, a quark can and does have all three types of color charge, but only one color at a time. In interaction with other quarks, a quark is a kaleidoscope that rotates in a rapid-fire way among the three colors. Perhaps U.S. politics could take a cue from quarks and diversify away from strict dichotomization.

Naming the types of quark "color" red, green, and blue might seem like more physicist zaniness, an inside joke. But there is some method to this nomenclatural madness, which we will see at the first event of combogenesis in the next chapter.

PRIMORDIAL THINGS AND RELATIONS

This book is about the progression of an ever-larger nesting of levels of things, from purely physical to biological and eventually to cultural. For that progression to happen, things at all levels have to have relations. And the quanta of the Standard Model do. For us and our quest for a unified narrative with potential rhythms from quarks to culture, here is what we gain from the division of the fundamental quanta into matter-field quanta (which quarks belong to) and force-field quanta: the universe started with a basic creation and apportionment of things and relations.

The New York University physicist Allen Mincer described the situation for me: the matter-field quanta toss force-field quanta back and forth. During this tossing, matter-field quanta interact and alter each other in various ways.[10]

Mincer said the kind of interaction we call "repulsion" is the easiest to understand metaphorically. You toss a heavy ball to someone, but you jerk backward a bit from the release, and so does the catcher, who absorbs the momentum of the ball. In the microcosm of the quanta, this tossing can create attraction or a moving closer as well as repulsion. That kind of interaction is more difficult to intuit. Perhaps it is like the tossing of a life preserver to a struggling swimmer at sea, by which two bodies can be moved closer together—it is hoped—to safety.

Which specific matter-field quanta toss which specific force-field quanta has to do with the kinds of properties possessed by the matter-field quanta. Quanta with color charge, such as quarks, toss gluons back and forth. Quanta with electric charge, such as quarks and electrons, toss photons back and forth. All the matter-field quanta have weak isospin, which allows them to toss the W and Z gauge bosons.

There is no perfect metaphor or even language for this situation that has been so well honed mathematically. The basic kinds of relations are commonly called the "fundamental forces." Many physicists now prefer to say that the force-field quanta "mediate" interactions among the matter-field quanta. Perhaps imagine rivers that carry sediments from mountains to oceans. Or a marriage counselor, minister, or judge negotiating between parties. Or bees pollinating flowers.

Whatever metaphors our minds conjure or the terms we use from physics, the big-picture point here is that despite what might seem to us humans as extremely nonstandard weirdness in the Standard Model, the universe did us a good turn. For us, the bottom line has to do with things and relations.

The situation is *not* as simple as matter-field quanta = things, force-field quanta = relations. For example, the force-field quanta called gluons also have color charges and can interact with themselves. Indeed, as Mincer described to me, all quanta are in some sense "things" that interact. They all have their properties that determine their interactions. I noted the complexities involving real and virtual particles. We likely bias our thinking about the situation because we tend to conceptualize structures in familiar see-and-touch kinds of way. Still, the matter-field quanta and force-field quanta are a basic division of types that is crucial to how the universe works.

■ ■ ■

At its primal birth, the stage of actors in our cosmos was rife with potential for higher-order structures. How simple it all began: a few basic mathematical properties seeded throughout a small array of types of entities. We can take the array of fundamental quanta to constitute the base level of the grand sequence (level 1, if you like).

The physicist Frank Close highlights what the fundamental quanta give us: "The electrons and quarks are like the letters of Nature's alphabet, the basic pieces from which all can be constructed."[11] Note the metaphor of the alphabet. The analogy is spot-on. In both of these systems—alphabet and quanta—simplicity generates complexity. I say more about this pattern throughout the book.

The grand sequence has begun. All subsequent levels to come trace their ancestry to the fundamental quanta of the Standard Model. Now all we have to do is to follow the story of how these primordial things and relations combined into larger structures.

This starter level is unique to what's coming because—in our scheme—its things do not derive from the combination and integration of smaller, prior things. Why not? Well, I did specify "in our scheme." In reality, there could have been a prior level of constituents that we haven't discovered and confirmed yet. Worldwide, top-notch physicists are volleying speculations about this very question.

Ideas and debates abound. One idea points to multidimensional strings from which distinct vibrational modes created the fundamental quanta. In yet another theoretical tack are hypotheses about entities called "preons." It is possible that solving any current mysteries will open doors to yet deeper ones. Or maybe not. And then there is the enigma, as yet unmentioned, of cosmic dark matter, probably an unknown class of fundamental particle. Let's not even start into cosmic dark energy.

Basically, what preceded this deep base level, if anything, is mysterious. Proposed answers are not yet ready to be presented in introductory textbooks. Therefore, they are not for this book. Best sellers have helped us nonphysicists glimpse the possibilities. But because I am not the one to speak on these matters, I invoke a phrase from philosopher Ludwig Wittgenstein, with apologies for slipping from the original context: "About what one cannot speak, one must remain silent."[12]

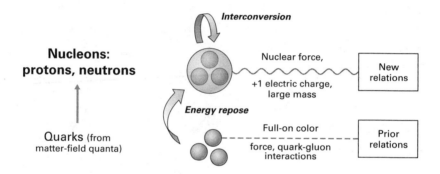

FIGURE 4.1

The figure above shows the new level, with its new things" (nucleons) and new "relations." Interconversion: inside stars, in radioactivity, and in other nuclear transmutations.

4

THE NUCLEONS, WITH IMMORTAL PROTON AND FRAGILE NEUTRON

SUMMARY: *Things that relate might integrate. Quarks and gluons in the fundamental quanta combine to create the first known systems. Quarks connect via the gluons into objects that balance the internal, superstrong color charges. Experiments have discovered a "particle zoo" of more than a hundred types of simple quark–gluon systems, with either doublets or triplets of quarks. Yet from that impressive diversity only two species of quark triplets emerge that are stable enough to become the real stuff from which we are made: protons and neutrons. Protons by themselves may well be immortal. But neutrons have a serious problem.*

THE NUCLEONS

Following the birth of the universe, billions of years passed before culture blossomed, at least on Earth. But at the first mere heartbeat of time, new things were suddenly born from combogenesis. At a new level, these things were the protons and neutrons.

Before the one-second mark after the Big Bang, quarks and gluons of the Standard Model lived in a maelstrom hell of hells called a "quark–gluon plasma." After that second, when the temperature of the expanding universe dropped below about a trillion degrees Kelvin, naked quarks and gluons were no more. They jelled into protons and neutrons. Physicists refer to this event as a "phase transition," akin to water congealing into ice. Because the average temperature of the cosmos kept on falling, this phase transition to protons and neutrons was like ice everlasting. A nonreversing ratchet to a next stage had locked in.[1]

At the end of chapter 3, I noted that it is *not* established that the fundamental quanta of the Standard Model are made of smaller things. Thus, we can honor the protons and neutrons of the level discussed here as the *first known systems* on the grand sequence.

Protons and neutrons are collectively called "nucleons." The word *nucleon* refers to the atomic nucleus, which jumps ahead in our story and shows the role the nucleons will soon play. But at the one-second mark of time, the nucleons emerged as things unto themselves, the most avant-garde "cool" items in existence, belonging, as we all do, to the greater cosmos but at this point not within larger structures.

The nucleons are therefore mighty special as the first systems. What nature of systems are they? And what capacities did they gain that were so novel that they could lead along the road of time up into a next event of combogenesis?

CREATING NUCLEONS FROM QUARKS AND GLUONS

To jell quarks into protons and neutrons by combination and integration, only two types of quarks from the suite of six are usually cited in the recipes: up quarks and down quarks. Each is respectively the lightest type (by mass) of the three quarks in its group.

I say "usually cited in the recipes" because other kinds of quarks enter the kitchen when we consider the bizarre goings-on inside protons and neutrons. But first we visit each recipe's sticky sauce. It's the gluon. Recall that in the prior level quarks are members of the class of matter-field quanta, and gluons are of the class known as force-field quanta.

The name "gluon" is apt among the often wacky monikers of particle physics. Gluons are what physicists call the "mediators of interaction" among things that possess color charge. Gluons are tossed back and forth by quarks during bonding of the quarks (and of the gluons, as discussed later). Such interaction manifests what is known as the "color force" or "strong nuclear force" or often just "strong force." The color force pulls things together. In a sense, gluons glue quarks together.

The color force among quarks and gluons is by far the most muscular of all interactions among the fundamental quanta and is a *main integrating factor* for this stage of combogenesis. It binds, and it binds with almost unbreakable heft. The internal dynamics of the first systems it creates are particularly tricky to visualize, for not only quarks possess color charges, but so do the gluons, which means they interact with themselves. Thus, the mediators mediate among themselves as well as among the quarks.

Why were nucleons the first systems to coalesce after the Big Bang? The keys are the heft of the color force and, as noted, temperature. High temperature is the enemy of stability in physical systems. High temperatures shake matter and can rip connections among parts. As the universe began its chilling after the Big Bang, the structures first born were those that coalesced via the strongest ligaments. Because quarks and gluons carry color charges, they were the chosen ones to combine. The systems that were formed enjoyed a kind of "energy repose," like a stretched rubber band allowed to relax.

Now for the recipes that create this repose within protons and neutrons. The simplest recipe cites two up quarks and one down quark in a proton, two down quarks and one up quark in a neutron. Gluons bind the quarks and themselves. The triplet of quarks in each nucleon relates to the fact that color charge comes in three hues.

Recall: red, green, and blue. Though these visual terms are metaphors, the language is more than wild fancy. In our normal world of visible light, when beams of red, green, and blue light merge, the result we see is white light. Deep down in the world of fundamental quanta, when a trio of quarks of three different colors are bound into a system, the colors are also balanced. Metaphorically, the result is white. In reality, it is similar to the balance of the energy repose between positive and negative electric charges in an atom (discussed later), but in the threes of color. In the words of physicist Martinus Veltman, "All bound states of quarks must have a color combination such that they are essentially 'white.'" The result, according to Veltman, is understanding a proton as a "glob of gluons with three quarks swimming in it." The "blobs of gluons . . . resemble wads of chewing gum."[2] The scintillating sinews of internal forces make these globs, blobs, or wads the most power packed in existence.

According to the Bible, God said, "Let there be light." Here, in our first systems, that metaphoric light was white.

THE ORIGIN OF MASS WITHOUT MASS

This simplified picture has been further elaborated by physicist Matt Strassler. Indeed, he calls the picture of a proton made of three quarks "a lie—a white lie, but a big one. In fact there are zillions of gluons, anti-quarks, and quarks in a proton. The standard shorthand, 'the proton is made from two up quarks and one down quark,' is really a statement that the proton has two more up quarks than up antiquarks, and one more down quark than down antiquarks. To make the glib shorthand correct you need to add the phrase 'plus zillions of gluons and zillions of quark–antiquark pairs.' "[3]

Where did these "zillions of gluons, antiquarks, and quarks" inside the proton come from? One day I naively wanted to compare the mass of a proton to the mass of its standard recipe of two up quarks and one down quark. I could just look up those three quark masses and then sum them up. Well, to my chagrin, the sum didn't come close to the proton's mass. In fact, they totaled to less than one percent of the proton's mass. My New York University colleague physicist Allen Mincer explained to my perplexed self that the huge amount of extra mass that was missing in my math comes from the internal quark–gluon and gluon–gluon interactions.[4]

Interactions? These interactions make a football game seem like a still pond. They are so intense that the maelstrom of tossing and shifting at nearly the speed of light within protons and neutrons creates and annihilates quark–antiquark pairs of virtual particles; quarks switch colors; pairs of up and antiup, down and antidown, strange and antistrange, even charm and anticharm quarks (in fact, all types) are born and disintegrate; and the color-possessing gluons also hinge and unhinge with each other and create and dissolve other gluons and quark–antiquark pairs.

All this interaction creates mass. Physicist Bruce Schumm invokes the formula $E = mc^2$ here. The "internal energy [of protons and neutrons] . . . will also manifest itself as mass."[5] Strassler calls the result "pandemonium."[6]

For us, the practical result of this interaction was most of the visible mass of the universe (there is more dark matter). In the protons and neutrons, mass is created from the unimaginable pandemonium of the internal things and relations bound into their nearly crazy-making systems. Thus, Schumm says that the "notion of mass" in the ordinary way we might think about it is a "sham."[7] The physicist Frank Wilczek describes this phenomenon as "mass without mass."[8] Strassler sums it up: "How remarkable is this!"[9]

DIVERSITY OF THE NUCLEONS

Because protons and neutrons will progress on to the next level of the grand sequence, they are the duo we need to know about at this level. What about other systems of quarks, though? Why just the proton and neutron?

Veltman said that the bound state of quarks must be "white." Indeed, other three-part systems of quarks can also yield a balanced a red–green–blue kaleidoscopic pandemonium and have been discovered. Quark duos consisting simply of a quark of one color and an antiquark of the anticolor are also a balanced "white." Such duos are indeed known to physicists.

In the 1950s and 1960s, new particles were being discovered so quickly in high-energy experiments, it was as if a zoo was being filled with new species. Physicists even called the bewildering variety the "particle zoo."[10] Research eventually revealed that most of these odd creatures were composites of quarks. Quark triplets other than protons and neutrons were being born—doublets, too.

Recall that the physicist Frank Close refers to quarks as members of "Nature's alphabet."[11] Theoretically, almost all types can combine. And in the zoo of color-balanced, "white" triplets, nearly eighty such composites have been teased into existence by high-energy physics. For quark doublets, about fifty species are verified and fit with theoretical predictions from the Standard Model. Such facts count among the model's many successes. Thus, well more than a hundred of these exotics have been found and given names such as "lambdas," "sigmas," "pions," and "kaons."

Experiments are ongoing and have recently reported unusual numbers, such as four and five, in quark systems, so there is news a-making.

But why haven't nonphysicists typically heard of these lambdas and their exotic brethren? Put succinctly, after creation they very rapidly go extinct. All the alternative triplets, when made, survive for only about a billionth of a second. And those are the long-lived ones. Thus, the vast majority of these triple-quark structures are bubbles in quantum fizz. They quickly burst apart in what is called "decay" into more stable forms of matter with energy that dissipates. The same goes for the doublets.

Fortunately for us, standing out from this diversity of fly-by-night particles, the proton appears to be a true immortal. Current supersymmetry theory puts the proton's lifetime at a minimum of more than a billion billion times the universe's known age.[12] The nickname "the god particle" was bestowed on the only recently confirmed Higgs boson, but what kind of god lasts only an ephemeral 10^{-22} seconds? Compared to the Higgs, the proton is a real god particle—a sacred, secure one.

We now turn to honor the second nucleon, the other of the two "mass-without-mass" foundations of all higher structures of matter. Geometrically, the neutron as a quark triplet seems not too very different from a proton, just one down quark rather than an up quark in the inner pandemonium of the zillions engaging with each other. But a neutron's lifetime make it seem fragile, like cosmic glass. A lone neutron, say in space, lasts only about the time it takes to drink a cup of coffee or scarf down a slice of pizza. After its fifteen-minute average shot to exist, a neutron "decays" into a proton, an electron, and an antineutrino, plus liberated energy that is carried away by the kinetics of those three physical products.

Though that brief candle of a lifetime is many billions of times longer than the lives of the other quark–gluon systems of the particle zoo (except for the godlike proton), that timeline is still not very hopeful for a world in which neutron-containing people like to live longer than it takes to down a slice. So wherever and whenever a neutron comes into existence—and some are being born in the cores of stars (including our sun), from protons (interconversion within the level), and also in certain kinds of atomic radioactive decay—for it to anchor itself securely into reality, something had better happen pronto.

THE NEW RELATIONS: NUCLEAR FORCES AND MORE

About the things of this new level, we can ask a question that will be repeated at every event of combogenesis: What *new relations* were born? Only with new relations will the new things created have the possibility to combine and make a subsequent next level that is itself, in turn, new. Several new properties were born with protons and neutrons relative to the previous level's quarks and gluons.

First, as covered, there was the creation of additional mass. Mass itself is not fundamentally new—quarks have smidgens of mass—but the boost in mass by a factor of more than a hundred times does facilitate the formation of stars and planets as giants that house subsequent events of combogenesis.

Second, with nucleons we now have things of known size. Diameters of protons and neutrons are a quadrillionth of a meter (10^{-15} meter). Thus, the nucleons are at least a thousand times larger than the maximum size of a quark (if not a mathematical point). So a proton or a neutron may be tiny, but at least it's a known something. Size had to begin somewhere. This level might be that place.

Third, protons possess an electric charge of +1 (from the net sum of two up quarks, each with +⅔ electric charge, plus one down quark, with −⅓ electric charge; gluons are electrically neutral). Using the same numbers, you can calculate that the neutron is electrically neutral (net of one up quark and two down quarks). So the summed total +1 electric charge of the proton is a new value of electric charge different from any of the quark electric charges. A new value of the same type of charge doesn't sound so fundamentally new by itself, yet this new value proves crucial in a subsequent event of combogenesis.

Finally, and perhaps most central for our story, both protons and neutrons emanate a weakened value of their internal binding color forces. Physicists call it a "residual" color force or residual strong force. It is about a hundred times weaker than the awesome forces inside the protons and neutrons[13] and is limited to very short distances around them. This emergent "nuclear force" comes to center stage in the next chapter. A limited residual color force doesn't sound very innovative. Yet its very feebleness will prove to be its creative strength at the next level. And it will help solve the challenge of the fifteen-minute natural lifetime of the fragile neutron.

■ ■ ■

The proton is in every atom of every cell in our bodies. It is in every atom of the universe. The proton is essential to the fabric of the universe. That is why there are plenty of lone protons out there even in the vastness of space among galaxies. But there are no lone neutrons because by itself the neutron is a particle with a short fuse and quickly disintegrates.

Note that what appears to be potential fireworks of types of quark–gluon systems in the Standard Model as base level has almost fizzled. Nearly all the hundred-plus quark–gluon systems found in the particle zoo of experimental physics vanish in flashes! Only the two special systems we call protons and neutrons remain, so to speak, to create possibilities for a level yet to come. Thus, how paltry seems the generativity of the Standard Model, this so-called periodic table of particle physics. Nature's alphabet has so far produced only two words. Contrast, for example, the well-known, super-rich generativity of the periodic table of chemical elements that create molecules. Contrast human language, whose small set of phonemes and letters allow a near infinity of words. This mere pair of stable products on this new level of physics is a wimpy result compared to the potential that the prior level of quarks seemed to have to form systems. This very sparseness gives one the odd sensation that this new level was just barely reached. Something in the universe just squeaked by, in some cosmic creative sense, in terms of giving us anything permanent.

But squeak by it did, and we now have our first systems from combogenesis: protons and neutrons. As new relations, their altered magnitudes of forces—electric and residual color—do not seem all that revolutionary. But we will see that key to the next event is a new relationship *between* these modified magnitudes, a relationship that cannot exist without this level.

5

ATOMIC NUCLEI FROM MUTUAL AID

SUMMARY: Once arrived, protons and neutrons are capable of linking by their nuclear forces. A marvelous mutual aid ensues. Together the nucleons combine to create on a new level new types of stable systems called "atomic nuclei." Protons, by cooperating with neutrons, can be bound with other protons into the new level's ball-like clusters instead of flying apart due to their repulsive electrical charges. And neutrons in those clusters, by cooperating with protons, are rescued from their inherent fragility. The main new relations that come with the creation of the atomic nuclei are the radiating fields from concentrated, positive electric charges, which vary across a digital spectrum of magnitudes. This innovation will prove key to creating the subsequent level.

CHALLENGES TO THE CREATION OF THIS LEVEL

All human bodies contain the nucleons—protons and neutrons—on level 2. But to get up to us, many more nested levels must be built. From the perspective of what we have seen in the narrative so far, down in the Lilliputian land of particle physics, what combogenesis will need to achieve to produce human culture seems almost unimaginable. But as the aphorism goes, "For a long journey, take a step at a time, and you are on the way."

As noted at the end of the previous chapter, in the set of the nucleons' new relations one relation is special. Each nucleon seals off from the

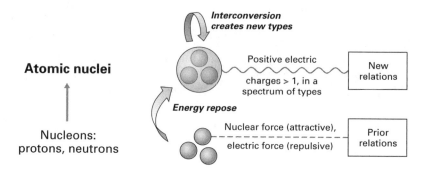

FIGURE 5.1

Atomic nuclei are composed of protons and neutrons, which first came into existence as types of things prior to the atomic nuclei. Interconversion: atomic nuclei interact with each other inside stars and during supernovas, which fills out the spectrum of natural types.

outside world the enormous, full-on color forces among quarks and their gluons. Well, mostly so but not completely. All nucleons have a residual color charge that they emit, called the "nuclear force," far weaker than the hundred-times stronger forces of the color-charged dynamos inside.

The nuclear force occurs because across the nearly kissing edges of nucleons, the internal color-possessing quarks of one nucleon are influenced by the nearest internal color-possessing quarks of the other. Thus, nucleons attract other nucleons when they are within a certain close range to each other. The nuclear force is the main factor of integration that will create new things at this next level.

In physics, where an attractive force exists between objects, the tension can be relaxed if the objects can potentially reduce their distance from each other. Such reduction of potential energy—an energy repose— is how gravity builds planets, stars, and galaxies. As a general principle, such reduction is how the full-strength color force builds up from quarks and gluons to the systems of protons and neutrons. The details differ in each of these cases, but the general principle of the relaxation of tension—potential energy—guides and orders various forces in play across scales. The principle works at this next event of combogenesis.

Thus, by way of the attractive nuclear force, an opportunity should exist. The opportunity would combine and integrate nucleons into new, larger systems. But there are problems. The situation is more complicated than colorful appeal.

Recall that protons also carry a positive electric charge with a new magnitude—namely, the +1 electric charge. Because the universe is structured such that things that carry electric charges of the same sign repel each other (positive repels positive; negative repels negative), protons electrically repel other protons.

Therefore, astir among protons are both an attractive force and a repulsive force. Indeed, in the opposing strengths of this simultaneous pull and push, the stickiness of the nuclear force is not sticky enough to stabilize protons exclusively into larger systems. Try to build such a system, and the explosively separatist electric force will dominate. Thus, no systems exist in which only protons are members, at least in the ordinary matter of our world and bodies.

Also, recall that neutrons have their own unique issue. Though they carry no electric charge, neutrons by themselves will disintegrate in about fifteen minutes, and the material by-products from this so-called neutron decay are blown apart from each other.

So in the task of combining prior things into new, larger things, the electric repulsive force among protons is one daunting hitch, and the neutron's brief lifetime is a second headache.

A BEAUTIFUL BINARY OF MUTUAL AID

Help fortunately comes via a remarkable subatomic cooperation.

Though a stable system of two protons cannot exist, tossing in one or more neutrons adds muted color glue to the rally but no extra electric repulsion. Again, that's because neutrons carry color but are electrically neutral. By adding neutrons to protons, we can get a stable structural system, a ball-like cluster of both.

For example, two protons and one neutron create the composite called a helium-3 nucleus. With two protons and two neutrons, the composite is a helium-4 nucleus. Helium-3 is extremely rare. But helium-4, though rare on a planet like Earth, is the second-most abundant nucleus out there in the universe, present inside stars and floating in the interstices of space, and its number is bested only by the hydrogen nucleus, most commonly a lone proton.

These simple, small, and new composite entities—such as helium-4, helium-3, and a smattering of a few others, such as deuterium, lithium, and boron—were formed at about the three-minute mark after the Big Bang. Astrophysicists have established that at that time a thermal threshold was crossed. Earlier, at the one-second mark, a threshold of cooling allowed quarks and gluons to stabilize into protons and neutrons. At the three-minute mark, the even cooler cosmos granted the ability for protons and neutrons to stabilize into the first atomic nuclei.[1] We called protons and neutrons the first systems known to be systems of parts, so here we might hail the nuclei as the first systems of systems.

But how was this event of combogenesis possible given the *pair* of problems that must be solved: protons repel other protons, and the neutron is so fragile and short-lived? As noted, the proton problem is solved by adding neutrons with color glue but no electrical charge. In addition to contributing nuclear force glue, neutrons also wedge the protons apart a bit in the balls, helping make peace by separating the antagonistic protons. But what about the problem of the neutron's intrinsically short lifetime?

Because many of the original nuclei—helium, deuterium, and so on—formed at the three-minute mark after the Big Bang are still floating around in the cosmos, we can deduce that being connected to protons must stabilize neutrons. With protons as neighbors, neutrons shift from mayflies to Methuselahs.

In short, having a proton as a neighbor prevents neutron "decay." When a lone neutron decays in an explosive blowout, one by-product is a proton. Were that neutron to attempt to decay while bonded to one or more protons in a community via the nuclear force, such a decay would "pop" a new, by-product proton into existence *right next to* an already existing proton. And that pop would necessarily bring into existence a huge amount of potential energy, given the popped proton's proximity to the original proton and the electric repulsion between protons. That new potential energy would be much more than available from the release of energy in a neutron's decay. Thus, the neutron's decay right next to an already-existing proton is prevented by a kind of thermodynamic back-pressure. It's somewhat like the way the floor on the second story of a house prevents you from falling downward to the first story.

For a neutron, the bonding to a proton by muted color glue is a kiss of immortality. And multiple protons bonded in a pack to neutrons benefit from the extra nuclear glue the neutrons contribute. With this remarkable dance of mutual aid, the new things of this level are enabled to come into existence.

AVENUES FOR NUCLEOSYNTHESIS

Atomic nuclei as integrated communities would seem to open a door to many sizes. The smallest of the communities would have only a few protons and neutrons. The larger ones would have lots of both.

In reality, however, the creation of a variety of atomic nuclei is not so simple as tossing neutrons into mixtures with protons. Protons and neutrons need to be forced close enough for the muted color forces to stick them together. Basically, that force comes from very hot temperatures and very high pressures. Conditions cannot be so extreme that the communities are unstable, but they must be extreme enough to force the togetherness inward across a threshold where the tight-fisted nuclear forces of attraction dominate.

Such sweet-spot conditions were met for smaller nuclei at the universe's three-minute mark. For most of the large number of classes of nuclei inside our bodies, sweet-spot conditions are also met in the centers of all stars you see in the nighttime sky. Inside stars like our sun, fusion reactions create nuclei of helium, carbon (six protons and six neutrons in its most common isotope), oxygen (eight protons and eight neutrons in its most common isotope), and a few other types of nuclei.

Crucial sites for kicking up the diversity of atomic nuclei are giant stars that go supernova. As a supernova develops and then explodes, its core turns into an alchemical cauldron within which particles smash into each other at ultrahigh temperatures and pressures. This maelstrom is a stupendous creative event. Such catastrophic star death explores all the ways for protons and neutrons to synthesize into communal balls of various sizes. Smaller nuclei get built into larger ones. Larger ones are split into smaller ones. This is interconversion within the level of the nuclei. And so it goes, back and forth, up and down the scales of possibility, forays large and small, nuclear reactions in tangled pathways.

Upon explosion, supernovas disperse their created products outward. The result is approximately eighty types of stable nuclei, a number based on how many protons each contains. For many of these types, the number of bound neutrons can vary (in which case the nuclei are called "isotopes"). Thus, depending on your criterion for the count, the full suite of stable types is nearly two hundred. The cosmic event also forges many unstable nuclei, usually those with the largest numbers of protons. Some of those radionuclides (in atoms, one level ahead) decay so slowly that they are with us today and contribute to minerals such as uranium ores and to the heat of Earth's core. Some radionuclides with us were derived from decay of other unstable radionuclides in Earth's past. Thus, chemists usually talk about ninety-two naturally occurring types of atomic nuclei (again, a count based on the number of protons in both stable and radioactive types).

Supernovas ancestral to the formation of our sun and Earth were the sources of matter for Earth and other planets of the solar system. We are the "stardust," the "billion-year-old carbon" that Joni Mitchell sang about.[2]

Compared to the mere pair of stable new things (just protons and neutrons) at the prior level, we see that the number of new things at this level has risen dramatically. With atomic nuclei, the grand sequence has gotten more complex, more productive. This level appears hopeful, promising.

A SPECTRUM OF TYPES OF NEW THINGS THAT RADIATE POSITIVE ELECTRIC CHARGES

What are the new relations born with the new things of this level?

To start, the color force is now at last truly closed off. Quantum color's ability to act as glue to integrate subatomic things to create this level is fully played out by reactions within the cores of ordinary stars and supernovas. After this level, pure electric charge will be the dominant type of relation in the larger things.

With atomic nuclei, the universe gains a spectrum of types. A digital spectrum of various atomic nuclei possessing positive electric charges comes about from the ability of the +1 electrically charged protons to be in clusters of various numbers with other protons, in mutual aid with neutrons.

The fact of this spectrum will prove crucial in the next two levels and is featured as the new relations shown in figure 5.1. In particular, the spectrum will ensure a similar wealth of diversity in the atoms of the next level.

We might here ask a question about these new relations similar in spirit to the question asked in chapter 4: Are these fields of positive electric charges really new or only new in magnitude? After all, the proton by itself on the prior level carries a positive electrical charge of one unit. What we have here is a variety of multiples of that single unit charge.

Indeed, one might claim that upward from the fundamental interactions of the Standard Model, there are no really new relations within or under the sun. This viewpoint might claim that what I have been calling new relations in the buildup of levels are, at least so far, modifications of those fundamental interactions.

I agree that the fundamental things and interactions of the Standard Model allow us to understand how everything else in the universe is possible. That base level with its fundamental quanta is the foundation. But we need to take a pragmatic approach to the issue of new relations. If a level of certain things came into existence only because things on a prior level had a certain kind of right stuff that did not exist two or more levels down, then that right stuff was significantly new enough. It was new enough to forge a new level of things that otherwise could not have existed.

Furthermore, what we are seeing here are not totally new charges and forces but new magnitudes of charges and forces that allow new balances to form, which could not have formed without those new magnitudes. Again, this is a pragmatic look based on what actually happened in our universe. Laying out the logic of this formation of ever-greater nesting is this book's goal.

■ ■ ■

The so-called fine-tuning here at this level of the grand sequence is superb. The strength of the nuclear force possessed by protons and neutrons and the other factors in play seem to strike just the right balance to stabi-lize a large variety of types of atomic nuclei, which together as a spectrum will feed into making that crucial (for us) chemical alphabet of atoms at the next level. Consider the words of physicists Stephen Hawking and Leonard Mlodinow: the "sum of the masses of the types of quarks that make up a proton," they say, "seem roughly optimized for the existence of the largest number of stable nuclei."[3]

This is not the place to visit the debates in fields from physics to theol-ogy about what this fine-tuning means. For now, we might be happy enough to contemplate it in awe—awe being a contributor to well-being and one of the pure pleasures of conscious existence.[4]

For a pattern is emerging. We have now seen two steps of combogen-esis: first, from the level of fundamental quanta to the level of nucleons and, second, from nucleons to this new level of atomic nuclei. In both cases, the new relations of the new things are modifications into new values of fundamental properties that originated at the deep base level of the fundamental quanta.

Projecting the pattern, we can expect that the nuclei and their new values of charges and thus forces will have the potential to participate in creating something innovative on a subsequent level. That must be the case because we exist.

6

ATOMS WITH SPACE-FILLING, ELECTRIC MANDALAS

SUMMARY: The grand sequence now reaches the atoms. In atoms, central, massive nuclei derived from the previous level tether electrons, mercurial things directly from the base level of fundamental quanta. The electrons orbit in space-filling, mandala-like communities of electric charges. This innovation gives atoms new kinds of relations: various degrees of surplus or lack of electrons relative to the preferred geometries of the electric mandalas. Such new powers will provide the means, via an atomic alphabet, to usher in the next step of combogenesis and an explosive field of possibilities.

INTEGRATING FACTORS FOR THE BIRTHS OF ATOMS

It's time for electrical forces to truly seize the day. Within the atomic nuclei created at the prior level, the quark–gluon color forces, whether full-strength or residual, have been sealed off, like ideal nuclear-reactor cores inside containment vessels. The positive electric fields emanating from those nuclei have now become the most powerful lassos of attraction around.

Yes, we must acknowledge the galaxy-molding, star- and planet-forming crunch of gravity. Gravity dominates on those gigantic scales. But particle to particle, gram per gram, electrical forces now rule at the microscales of combogenesis that we will watch on the next buildup in a sequence of ever-larger nestings.

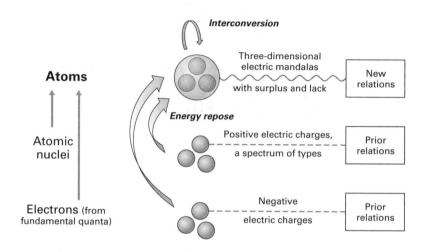

FIGURE 6.1

Atoms are composed of atomic nuclei from the prior level *and* electrons that have come "up," independently, from the base level of the fundamental quanta. The types of atoms correspond to the types of atomic nuclei. Interconversion: very weak, limited to nuclear changes (unless we count charged single-atom ions as types of atoms).

In addition to the positively charged atomic nuclei, key to this transition are negatively charged electrons. Say what? The creation of electrons goes back to the Standard Model that we started off with as the initial, base level of the grand sequence. And electrons have not yet been participants in the events of combogenesis. But they have been around, in a sense waiting, present on stage but not yet in a starring role.

The problem was that the conditions in which the first atomic nuclei formed (the early hot universe) were too disruptively hot for electrons to stabilize with those nuclei. The same goes for sites where nuclei are still forming (the interiors of stars, including the cauldrons of supernovas). But electrons were around in the early universe and throughout the bodies of stars, in plasmas with nuclei and photons, interacting but not integrating, like members of political parties too heated with antagonism to have stable conversations. And now we will see what happens when electrons and nuclei together reach cooler conditions.

As masses, electrons are relative lightweights. A single electron is only about 0.05 percent the mass of any individual proton or neutron. But an electron is a powerful package of negative electric charge. Its charge is exactly the same intensity as the charge of a proton but opposite in sign.

That means atomic nuclei and electrons attract forcefully. This attraction is the major integrator that can take the nuclei of the previous level and the electrons from two levels before that and unite them.

Whenever or wherever atomic nuclei and electrons were shifted into local conditions cool enough—again, the expanding early cosmos, ordinary stars, supernova stars—then those sites for the making of nuclei turned into sites for combogenesis and the creation of the varieties of atoms.

Specifically, in the sequence of steps after the Big Bang, after 400,000 years the temperature reached the next critical threshold. At that time, electrons and atomic nuclei could bond. This event created the very first stable atoms in our cosmos. The types of these first atoms were simply the atom versions of the same few simple types of atomic nuclei that had previously formed at the three-minute mark. To get the full spectrum of types of atoms in our bodies and world, we must look into stars.

The temperature of the cosmos at the 400,000-year mark was about the same as the temperature at the outer surface of a typical star like our sun. Edges of stars are generally cool enough for atoms to be stable (exactly where depends on the class of star). To get those atoms in other places as participants, some are blown outward by solar winds, away into space. But most importantly, when in their full spectrum of types atomic nuclei created in supernovas are blasted out into the cosmos, they reach cooler temperatures as they depart and pick up electrons that are all around in the gas bodies of stars. The electrons are exit prizes for the nuclei. The atoms for most of the full spectrum of the chemical elements we know and love of planet Earth came from past supernovas.

SPACE-FILLING ELECTRIC MANDALAS OF ATOMS

A lone proton from the level of the nucleon, two levels down, at cool enough temperatures will form with one electron an atom of hydrogen.

But a universe full of just hydrogen would be boring. Crucial for what is coming in the subsequent phases of the grand sequence is that atoms made on this level are generated in the same full spectrum of types as the atomic nuclei built in the prior level. In making electrically neutral atoms, those nuclei attract electrons equal to their number of protons (ignoring here the phenomena of charged ions and atomic bonding).

Thus, at this new level the universe gains atoms as communities of negatively charged electrons around central, positively charged nuclei. Atoms can have, to mention just a few of the ninety-two naturally occurring types of atoms, two electrons (helium), six electrons (carbon), eight electrons (oxygen), twenty-six electrons (iron), or many more, up to the ninety-two electrons in uranium atoms. Larger, highly unstable atoms with brief lifetimes have been made experimentally.

A remarkable feature of the general atom is its highly skewed distribution of internal mass. The nucleus holds more than 99.9 percent of that mass. The atom's emptiness is also renowned. To understand the scale of emptiness, picture the whole atom as enlarged to the volume of a classic European cathedral, in which case the massive but relatively tiny nucleus is about the size of a housefly at the cathedral's center. The physicist Frank Close stresses, however, that this so-called empty space is not the calm of a cathedral. Inside an atom, the volume of space, he says, is "filled with electric and magnetic force fields, so powerful that they would stop you in an instant if you tried to enter the atom."[1]

When electrons are bound around a nucleus, the force of electrical attraction tends to yank the electrons all the way down tight against the nucleus like a magnet to a refrigerator door. Complying with this tendency, even a little, tranquilizes the tension of potential energy inherent in the separation of the opposite electric charges of nucleus and electron. Here is yet another example of a general rule we have seen earlier: the minimization of separation when there are attractive forces. This rule or principle has been the main integrating factor in the events of combogenesis so far.

But were the electrons to fall all the way into the nucleus to maximally minimize that tension, it would be a disaster for our universe as we know it. Quantum mechanics has revealed how this complete fall is prevented. A trade-off between position and momentum, formalized in the math of the uncertainty principle in the late 1920s, has consequences for the stability and sizes of atoms. In essence, as the electrons get more confined by moving closer to their atomic nuclei in accord with the tendency to minimize the tension of separation, the momentum of those electrons becomes physically less determinate. The uncertainty principle is a reality crucial to the shaping of atoms. Thus, as electrons become more confined in space, they are freed to exhibit large fluctuations in momentum, which acts to scatter them away from the nucleus.

The resulting atom is thus a game of balance. Its size derives from a balance between contractive electrical forces tending toward collapse and the electrons' fluctuations of momentum tending toward expansion. The balance stabilizes electrons throughout an atomic space as relatively huge as the vast space of a cathedral is relative to a housefly at its center.

There are no walls to this metaphoric cathedral. Its real size extends indefinitely. But electrons have zones or clouds of more probable and less probable occurrence. Specifically, the flitting, quantum-jumping jerks and wavelike distributions of electrons in these communities with other electrons occupy geometric clouds of probability. Depending on the number of electrons in an atom's community, the atomic cathedral is filled with various geometric shapes: spheres, bulblike lobes, and partial doughnuts (toruses) called orbitals that are parts of broadly concentric zones called shells. These shapes, with fuzzy edges, are the clouds of probability within which one finds electrons at certain energy levels as the electrons make quantum dances in and out of existence within their given clouds. The mathematics that determine these shapes, their governing rules, was revealed by watershed breakthroughs from the same era as the uncertainty principle, and they include core concepts such as the Schrödinger wave equation and the Pauli exclusion principle.

To my mind, the patterns that result are wonderful and beautiful. They remind me of the contemplative mandalas of Tibetan Buddhism. These paintings display concentric rings of color and icons around a central core, which often focuses on a tiny *yab–yum*, a mating pose of male and female deities or other imagery such as a tight circle of three deadly passions, represented by linked animals. Though most commonly shown as two-dimensional paintings, the Tibetan mandalas can also be three-dimensional physical models or mental constructions built during rituals of focused meditation.

In the atom, the mandalas of electrons are what concern us here. Because the Schrödinger equation describes the wave mechanics of matter at the quantum scale, the mathematical beauty of the shapes of atomic orbitals and shells can be likened to wave patterns of music. Indeed, in contrast to the "pandemonium" or "wild dance party" of the interior of the proton and neutron, to physicist Matt Strassler the atom is "like an elegant minuet."[2] But it is the mandala metaphor that most grabs me. We might call the orderly arrangements of the communities of electrons in their probabilistic fields "the space-filling electric mandalas of atoms."

THE POSSIBILITY TO INTERWEAVE THE MANDALAS

In musical harmonies, you can be left feeling hanging just shy of a resolution, with an unplayed note that you hear in your mind's ear. Or there might seem to be a note too many, which also leaves you feeling unresolved. The same goes for the mandala-like geometric harmonies of the electron communities of atoms.

Most atoms have surpluses or lacks in the number of their electrons relative to certain magic numbers that constitute full shells.

Let's return to the analogy of the Tibetan painted mandalas. Suppose that tradition puts viewers most at ease when eight little painted Buddhas sit in a circle within the second concentric ring outside the mandala's center. But perhaps some paintings have in that ring only seven Buddhas. Perhaps some have nine. Observers would feel the need to remove one from the nine and add one to the seven. Perhaps one could be physically cut out from the nine and glued into the circle of seven, making eight in both. This is getting ahead a bit of the story of the grand sequence, but the analogy gives the essence of the idea of surplus and lack relative to some condition of energy repose even more relaxed than that of the basic atom itself.

This fact of surplus and lack of electrons is fundamental to the new relations born with this level's atoms.

In the atoms, as noted, certain magic numbers of electrons saturate the zones they can occupy. But most atoms are not exactly saturated. Their shells can have a deficit of one, two, or more electrons. And many types of atoms have a surplus of electrons, relative to the magic number of fullness. Because a saturated shell can slightly deepen the repose of energy by reaching a condition of even lower tension, these deficits and surpluses, relative to saturated numbers—eight Buddhas or eight electrons—entail the possibility that atoms can share with, take from, or give to other atoms.

With amounts of surplus and lack relative to certain geometries and numbers, we have the birth, here at this level in the grand sequence, of types of relations that will allow the electric mandalas to interweave in a variety of ways.

■ ■ ■

A patternologist might reflect on the scales of simple binary and trinary systems in all these levels so far: the electron–nucleus binary at this current level, the proton–neutron binary at the previous level, the quark–gluon binary and the quark triplets two levels down. These patterns are findings about nature's deep systems. They are not constructions primarily from the mind, such as the binary systems of yin–yang in ancient Chinese Daoism, the Tibetan *yab–yum*, or the sun-and-moon principles in the philosophic forays of the European Renaissance alchemists. These patterns are the way our universe is. Note the many binaries. Perhaps the trinary of quark colors is thrown in for variance to remind us about cosmic playfulness. OK, I joke a little.[3]

The binary system of electrons and the atomic nucleus creates relations from the new distribution of electric charges. Prior to the atom, electrons were not in coordinated communities. This condition preceded a coming complexity and subtlety of relationship. Nothing like this potential existed in the atomic nuclei. Those nuclei, once formed, could only repel each other. The nuclei could not remain themselves and also link with each other into things larger with truly different relations. But by combining nuclei and electrons, atoms did that.

The atoms even outdo the Tibetan mandalas, which might hang isolated on temple or museum walls or be stacked as separate scrolls on monastery shelves. With the atoms and their electric mandalas, we get forces more diversified and therefore more opportunities for connections more architecturally sophisticated than anything we have seen so far in the previous physical levels of the grand sequence.

Indeed, the vast possibilities brought into existence by the countable types of atoms are such that Trace Jordan and Neville Kallenbach, chemistry colleagues of mine at New York University, call the spectrum of types of atoms the "atomic alphabet."[4] The atomic alphabet opens up an expansive new space of potential for the generation of patterns as we move outward now to the next level.

7

AN EXPANDING CORNUCOPIA
OF MOLECULES

SUMMARY: Molecules are physical systems of atoms. For all practical purposes, the number of types of molecules is nearly infinite, given the rich interactions of the letters in the atomic alphabet. Molecules break out from what had been the general rule of spherical shapes of things on previous levels and possess new relations in virtue of electric charges across their bodies, which are electrical landscapes. With these innovations, molecules are capable of many kinds of bonding with other molecules and thereby participate in molecule-creating and molecule-destroying chemical reactions. Molecules, as a new level, open the potential to the origin of life.

MULTITUDINOUS COMBINATIONS

Some words from the days of ancient Rome still have a ring of truth. They were penned by a "Carl Sagan" of that period, a master explainer about the basic workings of our cosmos: "Multitudinous atoms, swept along in multitudinous courses in infinite time through mutual clashes and their own weight, have come together in every possible way and tested everything that could be formed by their combinations."[1]

The year was 40 C.E., the poet Titus Lucretius. From his masterwork, *The Nature of the Universe* or *The Nature of Things*, the Latin term *primordia rerum* has been translated as "atoms" in the quoted passage. Literally meaning "primordial things," the term has also been translated as "primary particles of things."[2] However the term is translated, Lucretius

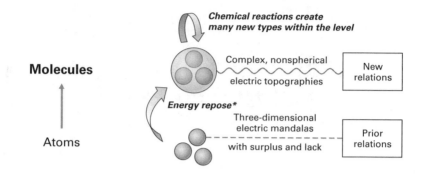

Molecules

Atoms

*Chemical reactions create
many new types within the level*

Complex, nonspherical
electric topographies

New
relations

Energy repose*

Three-dimensional
electric mandalas
- - - - - - - - - - -
with surplus and lack

Prior
relations

FIGURE 7.1

Molecules are composed of atoms from the prior level. Molecules interact with each other in chemical reactions to create (break and build) new molecules. (The asterisk after "energy repose" refers to the fact that this repose is a complex topic beyond this book because it involves comparison with other molecules in reactions: hence, the field of chemistry!) Most of the members of the vast set of molecule *types* are made within living cells through complex chemical reactions.

was promoting an ancient Greek philosophical idea of simple things creating complex combinations that we see in the world.

To be sure, Lucretius's atoms are not our modern atoms. And today physicists would point to the fundamental quanta of the Standard Model as the "primary particles of things." But our concept of atoms fits his poem because in the next sentence Lucretius says that the combinations of the *primordia rerum* form the "starting-point of substantial fabrics— earth and sea and sky and the races of living creatures."[3] Those "fabrics," as we now understand them, are directly made of molecules.

Molecules, of course, are combinations of atoms. But most molecules are not actually constructed directly from solo atoms. Yet the key to the molecules as combinations are the new relations that were born with atoms— namely, the electron communities with surpluses and lacks relative to saturated conditions of shells with preferred geometries. The three-dimensional, electric mandalas of atoms induced a wealth of prospects for linking into larger structures, given the right conditions. Extending this potential, molecules diversify by chemically reacting with other molecules, an ability at the heart of the new relations born with the things on this level.

Oh, but what about that "infinite time" Lucretius posited to realize the combinations? Astrophysicists currently say that it has taken 13.8 billion years since the Big Bang to get to us. Thus, if we cut Lucretius some slack, he did get it about right. It's been a long time of exploration.

MYRIAD TYPES OF MOLECULES

The first molecules hit the roadway of existence at about the 3- to 4-million-year mark after the Big Bang.[4] Why then? The basic reasoning is similar to what we saw for the first nucleons, atomic nuclei, and atoms, in that order. The temperature of the cosmos progressively fell. That decrease allowed increasingly weaker, new variants of physical forces of attraction, created by the events of combogenesis, to stabilize each next stage of bonding in which previous things linked into the new things. As temperatures dropped, the new, larger things then emerged in sequence with increasing sizes of expanding nestedness.

To make the first molecules, hydrogen atoms that were mutually clashing in the expanding cosmic stew joined into bonded pairs. This bonding took advantage of the fact that by sharing their single electrons, each one of two hydrogen atoms got a portion of two electrons, which saturated their shells, assuming they were willing to play nice and share. They were. (The preferred number for saturation is two in this primary shell of this simplest of atoms.) The result thus lowered energy tension from the common lack. The first molecule was H_2.

Now, when apart and solo, atoms repel each other because each one's negatively charged electric mandala mostly feels the other's negatively charged mandala. But if the atoms are pushed close enough together, they might actually bond. That is because their electrons also start to feel an attraction to the positively charged nucleus of the other, and the capacity might exist for some sharing or outright transfer of electrons to produce an overall result of energy repose in the new composite molecule system.

Atoms across the full spectrum of types are made and ejected from stars, primarily from supernovas, as described in the previous chapter. Then those atoms accumulate in cosmic gas clouds. Atoms collide. They merge into simple molecules of energy repose. Those molecules enter into

chemical reactions and make more types of molecules in more complicated states of energy repose. This level has a tremendous potential for
interconversion of types within it. We witness chemical dynamism in
such gas clouds by way of amazing color images taken by the Hubble
space telescope. For instance, astronomers have discerned stars being
born in the Orion nebula.

How many types of molecules are in those gas nebulae? Estimates are
derived using spectroscopes to analyze photons of light absorbed or emitted by the molecules. More than 140 types of molecules have been identified, including water, hydrogen cyanide, nitrous oxide, and ethanol.[5] Note
that these combinations of atoms already number more than the ninety-
two naturally occurring types of atoms themselves. This shows the
combinatorial potential of the atomic alphabet. Atoms are like letters.
Molecules are like words.[6] But in cosmic nebulae the potential vocabu-
lary is just getting started.

The Earth scientist Robert Hazen has investigated how many minerals were most likely present on Earth before life. Minerals are combinations of atoms in small crystals—for example, silicates, quartz, feldspar.
In a sense, minerals are giant molecules because groups of bonded atoms
are organized in repeating units (unit cells), which then connect on and
on over distances large enough to be visible to our eyes as tangible "fabrics." Hazen's estimate for Earth prior to life is about 1,500 types of minerals.[7] This pushes the number of molecules even higher than found in
cosmic nebulae. Words made from the atomic alphabet are getting more
complex and numerous.

But even that number is miniscule compared to the amazing potential
that this level holds—for example, the extraordinary number of molecular types inside the tiny, living cells of your body.

A gene of DNA (short for "deoxyribonucleic acid") codes for one or
more types of protein molecule made in a cell. A human has about 20,000
to 25,000 genes. That number implies at least that many types of protein
molecules in your cells (genes code for and correspond to proteins). In
addition, humans have more than 40,000 types of molecules called "me-
tabolites," such as various carbohydrates, vitamins, organic acids, lipids,
and other categories. Many of this vast array of metabolically crucial molecules are manufactured by proteins serving as enzymes or by proteins

and other molecules as systems, in groups sometimes called "nanomachines." Also, DNA and RNA (short for "ribonucleic acid") are molecules. Furthermore, the types of proteins vary in their details across biological species. A human's hemoglobin molecule in blood is different from a chimpanzee's.

So what's the overall diversity of proteins? Assume a number no smaller than the number of genes. Then how many genes are there in living things?

The Earth-system scientist and microbiologist Paul Falkowski addresses this question and concludes that a "reasonable estimate is probably on order of about 60 million to 100 million genes."[8] The estimate is even higher when we consider the total possible proteins in what biologists call "protein sequence space" (referring to the fact that proteins are large molecular complexes of smaller molecule units called "amino acids" in sequences): outnumbering the 100 billion stars in our galaxy. But there are issues involving facts on the ground, for these proteins have to function in life.[9] Falkowski's conservative estimate well proves the point: the number of types of molecules is huge, and all are made possible by the periodic table's relatively small number of kinds of atoms.

ODDBALLS WITH COMPLEX TOPOGRAPHIES OF ELECTRICAL FIELDS

In terms of shape, molecules are something of a breakaway level. As things, they literally are oddballs.

The entities of previous levels have shapewise been spheres or act as spheres. The quarks might be points or small balls (presumably) smaller than current limits of measurement can determine (about 10^{-18} meters). The protons and neutrons on the next level consist of a pandemonium of quarks and associated gluons whose ferocious, sinewy forces make the whole system oscillate in a blur so those nucleons form tiny spheres about 10^{-15} meters across. The smallest composite atomic nucleus on the next level is the deuteron, which is binary, with one proton and one neutron. But when the deuteron takes a role as the actual nucleus of an atom of deuterium, the single electron is so far away and the nucleus is spinning so fast in space that the nucleus acts like a sphere. As larger and larger atomic nuclei are made inside stars, the nuclei come closer and closer to being spherical, and, anyway, the nuclei function as if they were spheres in the centers of their electron communities when in atoms.

As for atoms themselves, typically 1 to 5×10^{-10} meters in diameter, it is true that some of their atomic mandalas of electron clouds contain lobes or partial doughnuts, say along x, y, and z axes. This internal partitioning of their essentially spherical presence does provide direction and structure for the atomic bonds they form with other atoms. But, again, when alone, as in outer space, atoms spin and vibrate and act as spheres. And atoms' internal partitioning comes into play precisely when they build into the nonspherical molecules and thereby contribute to that nonsphericity at a larger scale.

In contrast, molecules really break out from the sphere shape found on prior levels. Molecules are at minimum binary. That's not such a radical innovation for the case of H_2, in which both atoms are hydrogen. But when the two atoms are different, as in carbon monoxide (CO), the result is asymmetric, affording from its parts—one side a carbon atom and the other side an oxygen atom—asymmetric relations to other molecules in the environment.

Molecules, when spinning in space or bouncing around as gases in the atmosphere, can appear and act as spheres. But what's crucial about molecules is that they are not spheres. They have parts that form an architectural, nonspherical structure.

For example, consider the chlorophyll molecule. It looks somewhat like a tennis racquet with a really long and wobbly handle. Much larger is the Rubisco molecule. Rubisco is cited as the most abundant protein across all life. In its role as an enzyme inside cells of green leaves and algae, it is able to catalyze connections between carbon dioxide (CO_2) and a special three-carbon-atom molecule that cells possess. Rubisco brings CO_2 into living things. It consists of about 50,000 atoms of seven types (carbon, hydrogen, nitrogen, oxygen, phosphorus, sulfur, magnesium, and calcium).

A molecule of Rubisco appears roughly spherical. But crucial to its workings inside plant and algae cells is a zone into which a pair of smaller molecules can fit to be bonded together. In other words, the Rubisco molecule's *deviation* from a perfect sphere makes it function as Rubisco. The nook in the giant, bloblike Rubisco is not simply an open space like a cabinet but an inner surface replete with various electrical properties that attract or repel particular parts of the small molecules brought into the nook for catalytic mating. Giant enzymes abound in nature. For example, the double-blob cellulase molecule is excreted by the microscopic threads of certain species of fungi. Cellulase digests the cellulose molecules of wood and is thus crucial for the individual fungus and on a scale more mammoth for the global recycling of dead trees.

Molecules are tiny bunches of spheres (atoms) stuck together. A single grape (the atom) might be a sphere, but a cluster of grapes (the molecule) is definitely not. In living things, the nonsphericity of molecules is often related to their functions. The chlorophyll molecule's tail is anchored in a lipid membrane inside the green plant and algae cells it works within, whereas its tennis-racquet head, with a single large magnesium atom in the racquet's center, is designed to stick away from the membrane and serve as a solar collector that captures sun rays as they stream into the cells.

The variety of possible shapes found in molecules is enormous: structural webs, blobs, bridges, fantastic surrealistic shapes, scaffoldings, streamers, clouds, planes, tails, hooks, spikes, rings, and nooks and crannies of those protein molecules called enzymes, with their sites of catalytic action for performing operations of joining with and splitting from other molecules.

Crucially related to this generally new property of a molecule's nonsphericity is the complex topography of electrical fields emitted from a variety of electrical surfaces around the molecule body. Parts of the body interact uniquely with parts of the environment, which usually means other molecules but can mean atoms as well.

THE NEW RELATIONS MOLECULES CARRY

Thus, many molecules are incredibly complex, amazingly sophisticated landscapes of electrical charges. Unlike anything in the prior levels we have seen before, molecules have a playfulness in the "yes, we can" of their relations, especially when they start relating to each other.

We might pause here on humble H_2O, the water molecule. It looks somewhat like a round head (the oxygen atom) with two round ears (the hydrogen atoms) sticking up at two o'clock and ten o'clock, like a simple cartoon. It is polar. Each hydrogen atom's single electron is tugged toward a powerful electron grabber, the oxygen atom, which, when solo, lacks two electrons in its own mandala. Thus, in the overall system of the polar H_2O molecule, the two hydrogen ears are positively charged relative to the more negatively concentrated oxygen head.

This polarity gives the potential for weak bonds among water molecules. The hydrogen side of one molecule gets weakly linked to the oxygen side of other water molecules. These so-called hydrogen bonds among water molecules are like ribbons that form and break. Because the hydrogen bonds are weak compared to the tethering of electrons to nuclei in the atoms themselves, the hydrogen bonds play a crucial role: they make water, well, slippery, wet, watery.

I use the water molecule as an exemplar of the new, general relations of molecules. The new, wild freedom of shapes in the level of molecules is related to what is across their bodies: topographies of electrical charges of various magnitudes and even signs (positive or negative). Molecules create landscapes of electrical fields. This is like the way a mountain range has a topography of heights and passes even though the mountains are continuous and not isolated as individuals.

■ ■ ■

Consider a pattern here: a fairly small number of types of atoms gives rise to a nearly infinite number of types of molecules. This pattern, generalized, consists of a pair of sets of things. A small number of elementary types in the first set (atoms of the periodic table) leads to an astronomical generativity of more complex types in the second set (molecules). I have alluded to human language. A small, finite set of letters can form a nearly infinite number in the set of words. Atoms are like letters; molecules are like words.

There is a repeated pattern here. The pattern will prove important later, when we compare levels of the grand sequence. I call this pattern the *alphakit*. An alphakit consists at minimum of a set of "elements" and a set of productions huge in number and potential, like from a cornucopia. The alphakit will be prominent in other places of the grand sequence and will be essential to how we understand the overall story of the changing dynamics of things and relations when we have more of the sequence at hand.

The pattern of the alphakit is amazing. Specifically, ninety-two basic types of atoms enable the types of molecules to be, for practical purposes, nearly infinite. By "practical purposes," I mean there are types of sufficient complexity—a cornucopia of molecules—from gas clouds, Earth's geology, and living cells to enable the complex bodies of people to rise up. These people build sophisticated telescopes made of molecules that are tuned heavenward, and, via technologies made of molecules, they discuss findings about the types of molecules registered in deep space.

The complexity of relations among molecules, with their electric topographies and with the nearly infinite possibilities for creating and therefore connecting their types, sets the stage for the next level in the grand sequence: life. As a preview, note that most of the known types of molecules occur only within living things.

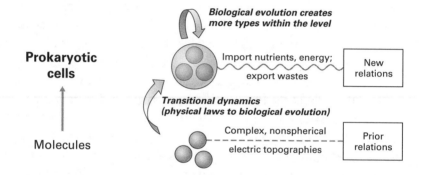

FIGURE 8.1

Prokaryotic cells are composed of molecules from the prior level. Cells have nested levels of complexity between the cell and its molecules, but the specifics of how the origin of the cell happened are contentious, and so proposed intermediate levels are not formalized here. The evolution and extinction of new lineages within the level are ongoing.

8

SIMPLE CELLS LAUNCH LIFE AND EVOLUTION

SUMMARY: This level brings us to the powerful innovations of the simplest living cells. This new thing, or being, is constituted— is a cooperative—of thousands to tens of thousands of types of interacting molecules. The cell is a nested biochemical system with numerous internal scales of living functions. This complexity entails revolutionary new relations: namely, cells require fluxes of nutrients as well as energy imports and waste exports. Being alive, cells grow and reproduce into offspring cells that in turn require their own imports and exports. As a consequence, living cells initiate the powerful, new dynamics called "biological evolution."

WHAT IS A CELL?

Molecules from the prior level naturally gather. They gather in cosmic nebulae of dust and gas. Some of those nebulae become stars, and around stars molecules gather into planets. On Earth, molecules gather into crystals, minerals, mountains, ice, oceans, air, clouds.

Molecules relate by their complex electric topographies. They have regions to and from which electrons can be exchanged or shared—in a sense, various modes of trade—generally with other molecules. At some point very early in Earth's history, molecules were apparently able to gather in ways to build the next new level of the grand sequence, the simplest living cells.

Modern biologists have reached a remarkable consensus along these lines: life originated from special, complex communities of molecules.

Inside a living cell, its system of molecules contains subsystems of mutual creation and destruction. Molecules help manufacture other molecules and are manufactured by others. Certain molecules serve as regulators of the cell's internal industries, dialing up and down rates of production of other functional molecules to maintain living conditions and to grow the organism. Some of the cell's molecules serve as borders for the commune as a whole, with controls on the gateways for materials that enter and leave. Some molecules serve as a genetic code for the patterns of other molecules made in the manufacturing nitty-gritty of ripping apart and putting together.

What happens in the community of molecules in a living cell is far more complicated and more sophisticated, from a systems perspective, than what goes on within the chemical reactions of atmosphere, ocean, or deep Earth. And it is in stark contrast to the molecules locked up in a crystal inside a rock, whose atoms can be neighbors for many millions of years without switching partners. We have not seen anything like this complexity in the previous five levels of the grand sequence.

From reading food labels, you know the general names for many vital types of molecules. In the previous chapter, these types helped exhibit the vast molecule spectrum: *lipids, carbohydrates, proteins*. Also deservedly famous is a cell's genetic molecule called DNA. So, too, are its smaller biochemical cousins, various types of RNA. Within all these main classes are subclasses and sub-subclasses. We might eventually probe into hemoglobin, cytochrome oxidase, nitrogenase, phospholipid. Going even finer grained in logic, specific molecules are unique to a biological species, such as the human hemoglobin molecule in contrast to the chimpanzee hemoglobin molecule.

THE BIOCHEMICAL ORIGIN OF LIFE

How the things at this level originated is a fascinating and still very much open question.

At the end of the previous chapter, I noted that most known types of molecules exist only in living cells. This fact makes the molecule–cell pair of levels unique among the levels we have seen so far. I revisit this fact for deeper consideration later in the book, after we have more levels as "data" that can contribute to the discussion. But for now this fact leads us to a version of the classic chicken-and-egg question: Which came first, the chicken or the egg? Our version takes place in what biologists and biochemists call the "biochemical origin of life."

Which came first: cells or the numerous types of special molecules that exist only within cells and that are produced only by the living metabolisms of cells? Both questions have the same answer for today's systems. Neither chicken nor egg is first; neither cell nor special molecule is first.

And yet when we turn to deep evolutionary history, we would give a nod of precedence to the eggs and the molecules. Before chickens, reptiles laid eggs, and those reptiles then evolved into birds that today lay eggs. Before cells, there had to be molecules, and there were. Moreover, these molecules had to interact in collective, semiclosed catalytic loops of mutual making that were cell-like in ways, even though those earliest ancestral chemical communities had to be simpler than those inside even the simplest cells alive today.

The chemist Addy Pross pegs the key shift here in whatever sites life was born as the transition from "ordinary chemistry" to "replicative chemistry."[1] Ordinary chemistry is what takes place in Earth's rocks, oceans, and atmosphere. In ordinary chemistry, molecules reach thermodynamic states of stability governed by energy repose, and they do so fairly predictably.

In contrast, replicative chemistry is both the logical and the real guts of life. In a cell, *chemical systems* propagate their patterns by the mutual cocreation of the special types of molecules that live, die, and grow as parts within chemically "replicative" communities. In a cell, the parts make the whole; the whole supports the unification of the parts.

There must have been a transition, says Pross, from ordinary chemistry of Earth to replicative chemistry that was a major shift, though we would not call the first forays fully alive. Life as we know it requires sophisticated replicative chemistry, with genetics or ribosome nanomachines for assembling amino acids into proteins, energy molecules that use phosphorus, cell membranes, and all else that cells universally have. But early on there must have been, so the logic goes, transitional biochemical steps. And at the final threshold to recognizable life, there emerged from this complex biochemistry the hypothesized LUCA, or "last universal common ancestor"—the ur-cell, progenitor of all current cells today.

The British biochemist Nick Lane also paints a scenario involving special kinds of chemicals, special chemical reactions, and special available sources of natural chemical energy. He shows that amino acids, the building blocks of proteins, would have been present in special, deep marine sites. So would simple forms of nucleic acids, the building blocks of the molecules that eventually were key to genetics. Lane narrates how these naturally constructed amino acids and chemical nucleic acids could have configured themselves into autocatalytic, self-making cycles that constructed, well, themselves in a progressive way in an early pathway toward life—all possibly within the tiny pores of minerals of alkaline vent chimneys deep in the dark, watery abyss of the ocean.[2]

These chimneys of "white smokers," says Lane, are the closest modern analogues for sites that could have given rise to life billions of years ago, probably during a time of unique conditions on Earth that do not exactly exist now. LUCA might have formed (evolved?) inside tiny porous mineral pockets of the smokers, which served as natural boundaries for confining the replicative chemistry. Early protocells might not have needed a formal border by way of membranes or walls, like those made by today's cells as functional skins. Perhaps protocells collectively created—as hundreds of interacting basic types?—what the biochemist Harold Morowitz has called a "core metabolism" of chemical reactions.[3] This core might possibly be found in all cells today as an inheritance and, moreover, a necessity, the heart of the origin of life and its subsequent expansion through evolution. Later in the process—or early, it's not known—some sort of genetics, as pattern storage and retrieval, allowed control over the kinds of molecules being made as these increasingly self-creating systems became more powerful at propagating their patterns.

As protocells became cells, cells eventually made their own sophisticated borders of molecules, which allowed them freedom from what had been the restricted though safe confines within the mineral borders of the pores of the white-smoker nurseries. The earliest cells could escape. They left the nurseries. They went wild. The early cell as a walled and membrane-bounded bag of self-propagating chemical reactions could start to live out in the wide ocean.

Pross likes to distinguish between specific scenarios, such as the white-smoker sites, and the overall logic.[4] He says that replicative chemistry must have been crucial in the buildup during the step of combogenesis from the level of molecules to the level of cells. The logic would apply even to scenarios involving panspermia. The term *panspermia* refers to the concept that Earth's original life came from elsewhere in space by some sort of cosmic seeding, intentional or not. The problem here from a theory standpoint is that panspermia merely puts the timing of the origin of life back into the distant pre-Earth past. The actual question about how cells emerged from communities of molecules would still hold. I personally don't invoke gods here.

In any case, fossil and geochemical evidence in sediments points to the production of cells perhaps around and likely prior to 3.5 billion years ago.[5] So long ago. We apparently are the descendants of a biochemical transition, a complex combogenesis from molecules to a first life called LUCA.

INTERMEDIATE LEVELS IN BETWEEN MOLECULES AND LIVING CELLS

In the sequence proposed so far, the molecule is a level used in creating the new level of the simplest living cells. Yet from the discussion just given, it is clear that many researchers who are focused on the question of the origin of life have posited a number of steps between molecules and cells.[6] Some of those steps might be considered events of combogenesis.

Cells contain complex nestings of parts, from giant, molecular nanomachines to tiny molecules, such as water. And nestedness in key molecules themselves is certainly crucial to cells today: millions of different protein molecules across species are made from the twenty basic types of smaller molecules called amino acids, and the genome is a large molecule of DNA that consists of numerous genes, then down to its bases, the famous small molecular As, Cs, Gs, and Ts of the genetic code. It is quite likely that these nested systems reflect ancient processes of combogenesis, perhaps as different types of chemical cycles of creation and destruction enlarged what was made by certain chemical pathways of combination. It is obvious to postulate the gulf between molecules and cells as having transitional "things," which combined by combogenesis and eventually led to cells.

But for me and my purpose, these postulates are not yet ready for the prime time of the grand sequence. The chemical loops of what Pross calls "dynamic kinetic stability"[7] would be candidates for an intermediate level. So are what Eörs Szathmáry calls "protocells."[8] But those two concepts, to cite just two, do not exactly jive. There is work to be done by others, not me. Szathmáry, citing progress, is confident that the science of the origin of life has moved "from 'unknown unknowns' to 'known unknowns.'"[9]

One issue will eventually have to be sorted out. Consider the human body with organs and tissues as "levels" in between the large body and small cells. These organ "levels" are not levels in the grand sequence, as will be clear in chapter 10 on multicellular organisms. The body did not derive from a combogenesis of prior-existing heart, lungs, brains, liver. These organs evolved *within the context* of the evolution of very much simpler, multicellular, whole creatures. Similarly, it seems likely to me that some of the nested complexity of cells came about ("chemically evolved"?) within a context of something cell-like. Nurseries for the origin of life in mineral pores of chimneys in the deep sea might have been the whole containers for the biochemical self-propagating loops and protocells, which might therefore themselves have been fairly large in size. To me, it seems almost certainly the case that some of the nested systems within the earliest cells that we would recognize as cells came about in the context of a very much simpler pre-LUCA system that might have progressed to life without actual combogenesis of independent, preexisting parts.

Fitting these considerations into the model of this book awaits more results from experiments and theory. At this point, I don't feel confident in assigning numbers to presumed levels between the molecule and cell, given the debates and darkness in this area of knowledge. I see the molecule as a level so complex in its forms that nestings of molecules can be considered systems of things and relations within that level.

NEW RELATIONS: IMPORTS OF NUTRIENTS, EXPORTS OF WASTES

Bottom line: a cell, once created, does require nutrient inputs of nutrients and energy as well as waste outputs. These inputs and outputs are revolutionary new relations.

We have not seen relations like this before in the grand sequence. Many molecules, such as the simple H_2 or CO_2, could exist by themselves in space. So could the atoms, say, of hydrogen, carbon, and oxygen. So could the nuclei within those atoms. So could the protons within those nuclei (recall, though, that neutrons would have the problem of a very short lifetime if alone in space). If quarks and gluons at the most fundamental level existed alone, it was only in the first microsecond of the universe. These physical systems from quarks to atoms to molecules, once made, do not require imports and exports to maintain themselves as things. (But complex molecules inside cells fall apart and need to be remade.)

The cell, however, is a different creature because it is literally a creature. One point to emphasize again is that many of the creature's molecules exist only because they are inside it and functional. The cell uses its imports and exports for the molecular manufacturing of its thousands to tens of thousands of special types of molecules.

To manufacture those fragile, expensive, complicated molecules, the cell needs imports of matter. By definition, these imports are what we call "nutrients." The molecules that serve as nutrient inputs constitute one group of molecular types that must regularly come in across a cell's border. And because actively building many of the cell's molecules requires energy, an import of available energy is needed. This energy can come in the form of chemicals from the environment that contain energy, which the cell can then metabolize. For cells that perform photosynthesis, the energy comes in electromagnetic radiation—in other words, light.

The manufacturing generates wastes. Wastes are another group of types of molecules that a cell must pass across its borders, but in the opposite direction, outward, ejecting them. Wastes are generated from the cell's internal manufacturing cycles in the way that human industries generate material wastes and as the human body generates wastes, for that yucky matter.

I need to get more specific here to set the stage for what is to come. In figure 8.1 at the beginning of this chapter, the type of thing—the earliest living cell—made from the combogenesis of molecules is the "prokaryotic cell."

The word *prokaryote* means "prekernel" or "before the kernel or nucleus." Accordingly, in prokaryotic cells alive today, for the most part the DNA is in a single chromosome that more or less freely floats within the cell. Thus, the genetic mechanics of this kind of cell are structurally configured differently from what we will see as the cell type at the next level. In short, prokaryotic cells lack a central membrane-bound nucleus. But what they bring to this level of the grand sequence are some major innovations in how these cells behave as systems, innovations that will greatly influence levels to come.

The innovation of cell replication through binary splitting was somehow a foundational process in the origin of life. In the most common way used by many of the simplest cells called "bacteria" and "archaea" (both prokaryotes), the replication usually occurs when the growth of the cells reaches a certain size. At that point, the cell divides into two smaller, "daughter" cells. The orchestrated cycle of growth and replication then repeats. Bacteria can also divide asymmetrically, and some species produce small buds of new cells that pinch off and start their own growth.

This innovation of replication seems to have been a consequence of the evolution of self-manufacturing chemical loops of living replicative chemistry. The ability to just maintain in the face of falling apart implies the ability to grow. A cell has to have the ability to grow molecules to balance the loss of others. "Net" growth results if the "gross" growth is more than the losses. Replication is a way of meeting challenges that such net growth presents. For example, when a given geometric shape gets larger its surface area increases at a lesser rate than its volume does. Thus, a cell that grew indefinitely would face a difficulty bringing nutrients to its interior with a *proportionally* smaller and smaller vital surface area. Replication solves that quite nicely by resetting the cell (now as two) back to a smaller size.

The hypothesized, intermediate levels of replicating, chemical loops referred to earlier also needed imports and exports. Thus, these revolutionary relations could have started at the lead-up to the origin of life. But no matter how many levels are posited in between molecules and the first cells, we know for sure that cells—among all the things that have preceded it in the grand sequence—have revolutionary new types of relations with their environments. LUCA lived by imports and exports.

THE TITANIC CONSEQUENCE: BIOLOGICAL EVOLUTION

There is much going on that is extraordinary in the cell in comparison to the molecules from which cells are built and to all the things of the previous levels—so much so that in figure 8.1 I have characterized this event of combogenesis as "transitional dynamics." The transition is from a "realm" of physical laws, with its many sequential events that have been governed by energy repose, to what are now the full-on dynamics of biological evolution.

Furthermore, the cell's innovative relations of imports and exports virtually assure, as I hope to show, the start of biological evolution.

The key to that nexus is a set of consequences from the imports and exports. Both fluxes cause changes in the environment outside any cell.

First, the import of nutrients to fuel a cell's metabolism depletes those same nutrients in the environment around the cell. Magnitude does not matter to the baseline logic here. Imports create depletion. That depletion is directly detrimental to the cell itself. Nutrients or energy or both have thereby become more scarce. Such depletion is also detrimental to other cells with similar metabolic needs that live nearby, many of which are often our cell's genetic kin born from ongoing cell replication.

Next, consider the cell's export of wastes. That flux causes an increase of those wastes in the environment. Because the waste molecules are by definition useless to that species of cell and are often internally toxic (the cell *needs* to shuttle them to the outside), the buildup of these wastes in the surroundings will tend to increase the potential of harm not only to the cell but also to other cells nearby, especially those cells, again, that are metabolically similar, to which the wastes are truly wastes. (Note, however, that to cells of different species, those wastes might be nutrients, but that is another story that involves evolution and is important in the combogenesis to the next level.)

Thus, both imports and exports cause inadvertent changes to the environment that are not healthy to the cell and its genetic kin. To put it bluntly, those changes increase the likelihood of death. Again, usually the concern is not just one cell but a population of cells, many of whom are generations of genetic relatives. So there is competition.

Cells regulate the construction of proteins (as noted, from certain smaller molecules called amino acids, which serve as building blocks, discussed further in part 3). Because of the complexity of genetics, mutations are understood to be inevitable. So lineages of cells from the same parent cell will eventually sport differences. This is not the place to summarize the many types of mutations. Important here is only the fact that to evolutionary biologists these differences result in variations in genetics across generations of cells. Lineages of related species in the tree of life can then evolve from successful new variants of cells: a creation of new types within the level.

Therefore, from the new relations—imports and exports from cells—and given what has been said about mutations, the classic "ingredients" for the process of evolution logically follow: (1) inherent growth and replication from the living cell's internal dynamics, necessary for the manufacturing of the complex molecules involved in life; (2) variations across the generations of the cells; and (3) competition among cells in a shared environment for gathering of nutrients, which decrease because of the imports, and for avoiding the increasing amounts of wastes from the exports.[10]

I again note that prior to the origin of life the postulated biochemical loops of replicative chemical communities and protocells required inputs and outputs and likely "evolved," in a progressive way. The loops and protocells that best replicated themselves would have "won" the contest to go ahead into the next generations of their patterns. This period has been called an era of chemical evolution.[11] In that hypothesized era before the origin of what we would call a living cell, the processes of propagation, variation, and selection would also have been in play.

■ ■ ■

Truthfully, the process described here is all pretty magical. Something major happened between molecules and cells, however many smaller steps might be postulated or eventually accepted as the case. I later call the start of biological evolution the "base level" of a new "dynamical realm" in the grand sequence. After this start, for the next few levels biological evolution will have to be invoked in the logic of the explanations of each event of combogenesis because the things in relationship—the organisms—are involved in evolutionary dynamics. As we have seen, those new dynamics were deep rooted from the new relations of imports and exports. Those relations continued because any larger things made of cells must also import and export.

Also, I noted a nestedness of certain molecules crucial to cells: large DNA molecules made of smaller molecules called bases and large protein molecules of smaller molecules called amino acids. What we have here is another instance of the alphakit pattern: a "genetic alphakit" within the complexity of types of molecules. This general alphakit pattern will grow in importance over the grand sequence, as I hope to show as we progress.

GRAPHIC CONCEPT SUMMARY

———

The following insert is a gallery of examples from the fundamental levels. I hope you will periodically refer back to these examples as you read through the chapters to review how each level works with the others in the grand sequence. Though all twelve fundamental levels currently exist as a nested set of systems essential to our lives, each came into existence one by one, at its own time. Previously formed or evolved things combined and integrated into new things, and new relations or generative powers came about that enabled each subsequent level.

As a quick summary, the levels are:

1. Fundamental quanta (chapter 3; *start of the realm of physical laws*)
2. Nucleons: protons and neutrons (chapter 4)
3. Atomic nuclei (chapter 5)
4. Atoms (chapter 6)
5. Molecules (chapter 7)
6. Prokaryotic cells (chapter 8; *start of the realm of biological evolution*)
7. Eukaryotic cells (chapter 9)
8. Complex multicellular organisms (chapter 10)
9. Animal social groups (chapter 11)
10. Human tribal metagroups (chapter 12; *start of the realm of cultural evolution*)
11. Agrovillages (chapter 13)
12. Geopolitical states (chapter 14)

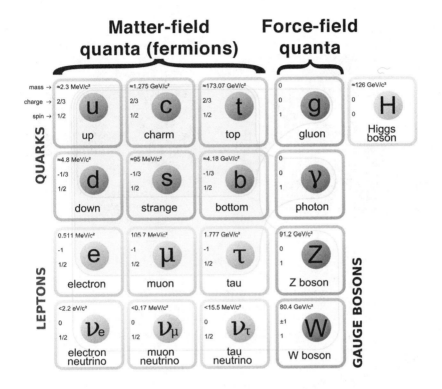

FUNDAMENTAL QUANTA

The Standard Model of particle physics, the base level of the realm of physical laws and therefore of the entire grand sequence. The quarks and leptons together are members of the fermions, the group called the "matter-field quanta" in this book, and the gauge bosons are the group called the "force-field quanta," following the terminology in Schumm 2004. The Higgs boson, associated with the Higgs field that imparts mass to mass-possessing fundamental quanta, is different from either of those two groups.

NUCLEONS: PROTONS AND NEUTRONS

The proton (*left*) and neutron (*right*). Their insides are roiling fluxes of energy, with multitudes of gluons, quarks, and antiquarks (with flat hats) that come in and out of existence (u = up quark; d = down quark; s = strange quark; c = charm quark; g = gluon). The quarks and antiquarks are balanced, except for an excess of two up quarks and one down quark for the proton and one up quark and two down quarks for the neutron. Those net "valence" quarks are highlighted for visual convenience, but they are not special. Diameters of protons and neutrons: one-trillionth of a millimeter.

Source: Adapted from Randall 2011

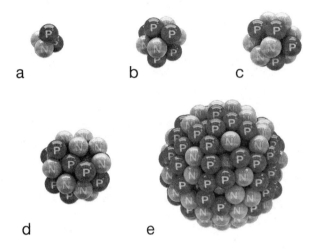

a   b   c

d   e

ATOMIC NUCLEI

The atomic nucleus is composed of protons and neutrons. Like the protons and neutrons themselves, an atomic nucleus is not a static structure as diagrams like these almost invariably show. It's impossible to depict the colossal forces and blinding fluctuations at work in these systems. Shown here (atomic number = number of protons): (a) helium (2); (b) carbon (6); (c) oxygen (8); (d) phosphorus (15); and (e) platinum (78). Diameters range from about one- to ten-trillionths of a millimeter, depending on numbers of protons and neutrons in the nucleus.

Source: general-fmv/Shutterstock.com

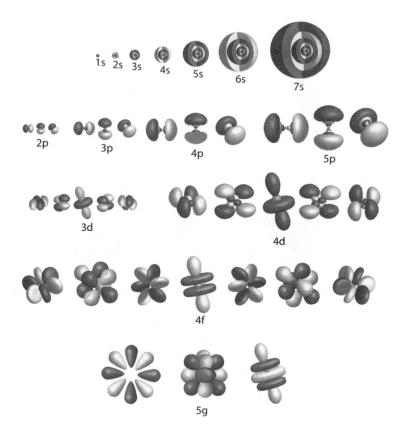

ATOMS

Electron orbitals of atoms. The massive but tiny nucleus is invisible at this scale. In reality, the electrons occupy clouds of probability; thus, the orbitals do not have hard edges but fade in intensity outward. Depending on its atomic number, an atom combines specific electron orbitals into a complex, three-dimensional "electric mandala," with families of orbitals that vary in configurations and sizes. The orbitals would be superimposed, with the atomic nucleus in the center of the collective mandala. Groups of orbitals with the same first number make up the "shells"; thus, shell "two" is made of the 2s and 2p orbitals. Diameters of atoms are not hard sizes, and the diameter of the bulk of the probability distribution depends on the chemical element, but atoms are about 30,000 times larger than their nuclei, therefore about 0.03 to 0.3 millionths of a millimeter.

Source: magnetix/Shutterstock.com

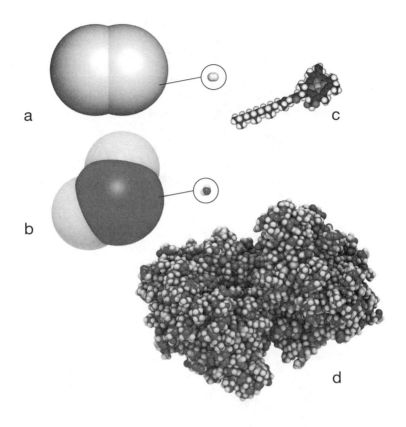

MOLECULES

A suite of molecules: (a) H_2, the first molecule in the universe; (b) thirst-quenching H_2O; (c) chlorophyll, which captures sunlight for plants and algae and makes the world green; and (d) a cellulase enzyme of several thousand atoms, made by a species of soft-rot fungus to break down cellulose molecules in the dead wood that the fungus feeds on. Sizes of the molecules are displayed so that the atoms are the same scale. The H_2 and H_2O molecules are shown enlarged as well.

Source: molekuul_be/Shutterstock.com

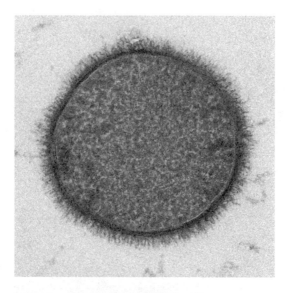

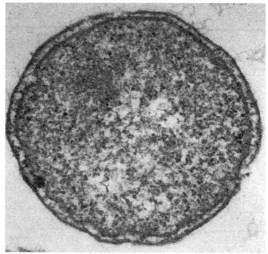

PROKARYOTIC CELLS

Top: Cross-section of the rod-shaped bacterium cell *Bacillus subtilis*. *Bottom*: A member of the prokaryote group called *archaea*: a large *Ignecoccus hospitalis* cell. *B. subtilis* is found in soil and even human guts. *I. hospitalis* was recently discovered in deepsea vents (they like it hot). Diameters: *B. subtilis*, 0.6 microns or 0.6 thousandths of a millimeter (typical rod lengths are ten times larger); *I. hospitalis*, 2 microns.

Source: (*top*) Allon Weiner, Weizmann Insitute of Science; (*bottom*) Reinhard et al. 2002

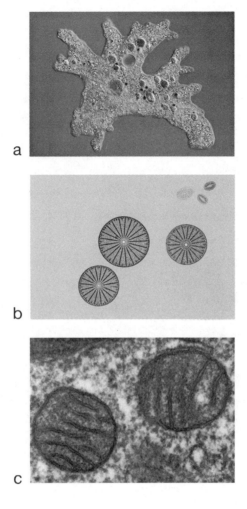

a

b

c

EUKARYOTIC CELLS AND THEIR INTERIOR MITOCHONDRIA

(a) The giant, free-living eukaryotic cell is the protist *Amoeba proteus*. (b) The round diatom is *Arachnoidiscus*, a marine algae, with its geometric shell of silica (akin to glass). As eukaryotic cells, both the amoeba and the diatom contain mitochondria, evolved from the common ancestral endosymbiosis. An amoeba will have thousands of mitochondria, slightly larger than the mitochondria of a lung's cells but still invisible on this scale; they need to be specially stained to be seen. Shown also are (c) a pair of mitochondria from a mammalian lung cell. Sizes: amoeba, about 0.3 millimeters; diatom, diameter about 0.2 millimeters; mitochondrion, 0.5 microns, therefore 0.0005 millimeters.

Source: (*top*) Lebendkulturen.de/Shutterstock.com;
(middle) Jubal Harshaw/Shutterstock.com; (*bottom*) Louisa Howard

MULTICELLULAR ORGANISMS

Shown here: bobcat (*Lynx rufus californicus*); barred owl (*Strix varia*); marine barrel sponge (genus *Xestospongia*), one of the world's simplest animals in terms of number of different cell types; oak tree (genus *Quercus*); honey fungus mushrooms (genus *Armillaria*), the fruiting, reproductive bodies of the underground network of fungal hyphae of the entire organism. The probing eyes and ears of the bobcat and owl highlight the new relations that senses provide for certain creatures on this level by using vibrations of light or sound.

Source: bobcat: Matt Knoth/Shutterstock.com; owl: Jill Lang/Shutterstock.com; sea sponge: Lawrence Cruciana/
Shutterstock.com; oak tree: Max Sudakov/Shutterstock.com; mushroom: Alex Coan/Shutterstock.com.

ANIMAL SOCIAL GROUPS

Shown here: a group of bonobos (*Pan paniscus*) in Africa; a pod of spinner dolphins (*Stenella longirostris*) in the southern Red Sea, Egypt; a swarm of bees on a tree (insects have evolved the sophisticated, eusocial lifestyle many times); red-spotted siphonophore (*Forskalia edwardsi*) in the Mediterranean sea, which is a colony of animals (zooids) that are functionally specialized and differentiated from clones that stay together and so is a very close-knit social group.

Source: bonobos: Sergey Uryadnikov/Shutterstock.com; dolphins: timsimages/Shutterstock.com; bees: A.G.A/Shutterstock.com; siphonophore: Seaphotoart/Shutterstock.com

HUMAN TRIBAL METAGROUPS

Here represented by rock art (the "Gwion" or "Bradshaw" rock paintings) created by ancient aboriginal hunter–gatherers in western Australia. "Although dating these figures remains elusive, a single tantalizing optically stimulated luminescence date of sand grains from a mud wasp nest that overlies 'Gwion-style' paintings suggest a minimum age of 17,500 years for the art work" (Donaldson 2012:1001).

Source: TimJN1/Bradshaw Art

AGROVILLAGES

Rice terraces in the mountains of the Philippines, Bangaan village, Ifugao region. The fields, as living components of the extended "village," function as solar collectors for nutritious matter and energy.

Source: Photomaxx/Shutterstock.com

GEOPOLITICAL STATES

This level is represented by the ruins of Monte Albán in the Oaxaca region of Mexico. This city once thrived as the capital of the Zapotec state, which formed about 2,000 years ago and was able to expand control over an area of about 20,000 square kilometers.

Source: Bo-deh~commonswiki

9

THE SEXY EUKARYOTIC CELL

SUMMARY: Evolutionary biologists hail the origin of a new kind of living cell that evolved about two billion years ago. Reaching a new level, this "eukaryotic cell" comes from profound biological symbiosis: a combination and then integration of a pair of different species of prior, simpler types of cells. One side of the pair evolves into specialized power organs inside the new cell. New relations that will prove consequential include true sex and an amplified potential to evolve sophisticated abilities to cooperate with other cells of the same species. The creation of this larger (on average) and more complex new cell leads to various evolutionary lineages that are later to evolve into multicellular fungi, plants, and animals.

PROKARYOTIC TO EUKARYOTIC: A NEW LEVEL OF CELL, A SPECIAL SYMBIOSIS

Once the first, simplest living cells came into being at the prior level, the engines of evolution revved. In Earth's oceans, those ancient cells related to each other in rich ways. They all took in nutrients. They all exuded wastes. Thus, they affected their shared chemical environments and each other.

Evolving, some species of cells discovered the means to utilize the wastes from others as nutrients for themselves. They established food webs based on these wastes, on other nutrients, and on parasitism, attack, and defense among various species. Given what we know about modern-day

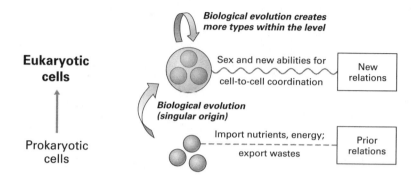

FIGURE 9.1

Eukaryotic cells evolved from a symbiosis of the two types of prokaryotic cells, bacteria and archaea. Other changes, such as the origin of the cell nucleus, are more mysterious. Biological evolution is operating; thus, the evolution and extinction of new lineages within the level are ongoing. I note what is currently thought to be a singular origin for this major evolutionary transition.

bacteria, some ancient species likely evolved quorum sensing. Here, individual cells switch behaviors, triggered by signals from others. Virtually from the start of life, modes of both competition and cooperation played off each other in ways that became ever more sophisticated and interwoven within the ancient microbial ecosystems.

One important form of cooperation is symbiosis: living together in mutual benefit. A familiar example is a flat green lichen on the surface of a rock. In this case, the symbiosis is typically between one or more species of algae and a species (or two) of fungus. In the lichen, the algal cells provide food via photosynthesis. The fungal cells provide fibrous structure for the shared home base. Another example of symbiosis involves microbes inside the bodies of termites. The microbes are allowed residence in the termites' guts and in return are crucial in helping termites digest the tough cellulose fibers of wood. Multiple species dine as one.

In one ancestral case of symbiosis, the combination became a new being worthy of a new level. Along our grand sequence, this new level is hailed by many biologists as a landmark of evolution second only to the origin of life itself. It is one of the "major transitions of evolution."[1] As far as we know, without this transition there would be no fungi, no plants, no animals. This level spawned stellar consequences.

The symbiosis that created this event of combogenesis involved more ancient types of cells that became a new type of cell. This integration event was so significant that biologists have given the new type of cell a special, big-picture name. Cells from the prior level were the "prokaryotic" cells, introduced in the previous chapter. The new cells are the "eukaryotic" cells.

Eukaryotic, prokaryotic—frankly, these words are difficult to love. For one, they are too close to each other in sound. I thought about calling the new cell type "the complex cell." But that's unfair to bacteria, which, as prokaryotic cells, are not at all biochemically "simple," although they are smaller and simpler when seen through a microscope. And so let us use the established biological terms, and the best way to remember them is to understand their meaning.

The most visible, defining feature of a eukaryotic cell is its nucleus, a prominent blob within the larger blob of the cell. The nucleus provided inspiration for the term *eukaryotic*, from Greek, meaning "good kernel." (We will see the same prefix *eu-* later in those superbly communal insects, such as ants and colonial bees, that are called "eusocial insects.") Thus, the word *eukaryotic* focuses attention on this cell's distinctive, membrane-bound nucleus, which houses the chromosomal DNA and other functions of the cell's genetic mechanisms.

Recall that, in contrast, the prokaryotic cell, "prenucleus," lacks a nucleus. This brings up another reason I'm not fond of the terms. Prokaryotic cells are thus defined by what they don't have, as if they are somehow lacking. It's a bit like watching a dramatic sunset on rosy-dappled snow-tipped mountains and exclaiming, "Isn't the prehuman beautiful!" As a class, the prokaryotes have been thriving a long time—basically since the origin of life. LUCA, the primal ancestor of all, was presumably a prokaryote.

For where we are headed—how a symbiosis of prokaryotes made a eukaryote—the fact of two main groups of prokaryotic cells is crucial.[2] One group is the bacteria, as noted. The other group is the archaea (with some debate over terms for both groups, but *bacteria* and *archaea* are most common).

Each group contains many species. For the most part, the members of each (imagine looking through a microscope) are as internally bland as plain porridge. Also, members of each are really small (as a rule today, but giant exceptions exist). The two-tribe reality of bacteria and archaea was revealed by genetics. Indeed, "bacteria versus archaea" is *more* unique as a pair of broad genetic classes than, say, "plants versus animals." Furthermore, the evolutionary split between bacteria and archaea is very likely rooted way back almost to the common origin of life. Their cell walls differ in chemistry and structure. Like the breakout of two tech companies that have hit on roughly the same concept but with distinct designs, perhaps innovations in borders allowed bacteria and archaea cells to break out as primal types from their nurseries in deep-sea vents and to start to live free in the global ocean. It might even have been a "live free or die" situation back then.

I dwell on these two groups of microscopic prokaryotes because the merger that created the eukaryotic cell apparently utilized species of *both* bacteria and archaea. When? The date is being debated. But, give or take about 500 million years, it is likely to have happened about 2 billion years ago, approximately halfway in life's tenure on Earth.[3]

ENDOSYMBIOSIS: ONE CELL LIVES INSIDE THE OTHER

Here is the essence of the symbiosis as currently understood. Probably a small population of a species of bacteria, as the smaller symbiont, took up residence inside a larger-bodied archaean cell as host. This type of coming together is called "endosymbiosis" because one member lives inside the other.

At first, like a centaur that is half-human, half-horse, the endosymbiosis was a chimera. The cell-like thing was part bacteria, part archaea. But then evolution took the symbiosis even further to create what evolutionary biologists designate as a "major transition." The entity evolved into an individual. What does that mean? Well, the parts became totally dependent. They evolved to reproduce together as a single cell. That the ancestry of the eukaryotic cell was thus forged entered the textbook phase of knowledge only after the biologist Lynn Margulis did her pioneering work in the 1960 and 1970s.[4]

What drove this step? Exactly what benefit of living together was brought to these two types of cells in that deep evolutionary past is still being figured out. Some researchers even claim that the symbiont began as a parasite inside the host. There is also talk of an initial enslavement or perhaps occupation, depending on who—symbiont or host—is argued as having the upper hand. Microbiologist Paul Falkowski thinks that the symbiont, by residing inside the host, got first dibs on wastes from the host and then in turn created its own wastes, which were then available to the host. Thus, the initial advantage was an arrangement of recycling.[5] Many biologists think some kind of interdependency and mutual benefit like this, even if not the initial cell bond, was the ultimate driver to a profound partnership. The bacterial and archaeal species were to share the same bed forever. In fact, one partner was the (living) bed.

Host and symbiont species presumably explored myriad experiments. How great it would be to have a time machine to go back and see what was happening a couple billion years ago. Biological evolution was working. Now, mutual dependence itself is fairly common in highly beneficial symbioses, as in lichens and termites. In the particular case of concern here, the cell endosymbiosis evolved into a full integration into that unified, new type of individual. Like the start, the "how" is still being debated, but on this question lots of progress has been made. First, what happened to the bacterium during this event of integration?

The bacterium—in a coevolution with the archaean host—became a crucial subsystem, a type of cell organelle, within the new cell.[6] In today's eukaryotic cells, these organelles are called "mitochondria." The number of tiny mitochondria, often shaped like human kidneys, inside eukaryotic cells can vary from one in some species to thousands in others, as in human liver cells. (The topic of eukaryotic cells that have lost mitochondria does not concern us here.)

The mitochondria are often called the "power plants" of eukaryotic cells. The internal mitochondria use nutrient molecules that the whole cell (as operational system) brings to them. The mitochondria then convert those nutrient molecules into special, high-energy small molecules (adenosine triphosphate, or ATP) that are distributed as the energy currency to places where they are needed in the eukaryotic cell.

It makes sense that mitochondria are so abundant in liver cells because our livers are the organs where nutrient molecules from digestion are processed into modified chemicals targeted for distribution to the body's other cells. Thus, the liver, as energy transformer for the body, has cells filled with organelles that are themselves like the "livers" for each eukaryotic cell.

The mitochondria are therefore like refineries that turn out a very special kind of fuel—say, a special jet fuel—for the vast metabolic, replicative biochemistry that is the living eukaryotic cell. Mitochondria make their cells' metabolisms fly.

A CELLULAR ENERGY REVOLUTION

During ancestry of the eukaryotic cell, many other evolutionary changes also happened. The cell's visually defining part, the nucleus, at some point "happened." When? How? Those questions are hotly debated. In addition, a generic eukaryotic cell has internal networks of trusses and wires that provide structure and serve as transport networks for metabolic materials. Other internal, membrane-bound organelles evolved as well. Notably, a eukaryotic cell can import nutrients as large particles, which can even be other cells it feeds on, like a dog able to wolf down big chunks of food. Some biologists have championed this phagocytosis, or "ingestion of other cells," as the key early step in the evolution of the proto-eukaryotic cell,[7] although others champion the nucleus, and others, as we shall see next, champion the mitochondrion. The science has a ways to go! Very exciting!

What role did the bacterium that evolved into the mitochondrion play? The mitochondrion is the vital power-plant organelle of the eukaryotic cell—that fact has been known for some time. Moreover, the biochemist Nick Lane and the biologist Bill Martin have made the point that the mitochondrion was the evolutionary innovation that allowed the new type to become a new level of the grand sequence, the quintessence of the event.[8] And that innovation has something to do with energy.

The more ancient prokaryotic cell (as a type, whether bacteria or archaea) was prevented from getting larger and more complex in the ways the eukaryotic cell eventually did. For the prokaryotes, those tempting evolutionary directions of size and complexity would have required more genes to control larger, complex metabolisms. But the eukaryotic cells, with distributed numbers of mitochondria, achieved a revolution in energy flux and control. Lane and Martin estimate the energy available per gene for the eukaryotic cell rocketed up by a factor of thousands, tens of thousands, even hundreds of thousands of times.

Regarding body size, ranges do overlap for modern prokaryotic and eukaryotic cells. That overlap is not unexpected. Evolution can produce outliers within all classes of living things. (My favorite "law" in biology is that there will always be exceptions.) But, on average, compared to prokaryotes, eukaryotic cells have about 10,000 times more volume to house their suite of distinctive internal structures and functions.[9]

Lane further points out that the other innovations of the eukaryotic cell, some noted earlier in this chapter, were also explored in sketch versions by various species of bacteria.[10] But it was the organelle of the mitochondrion, Lane and Martin claim, that, once evolved, made all the other innovations truly powerful in the eukaryotic cell and finally able to be incorporated as a suite into the cell's physiology.[11] No sense building wings big enough for a jet airline without yet having a jet engine to provide power.

THE NEW RELATIONS DERIVE FROM SOLVING GENETIC CONFLICTS OF INTEREST

The innovative, new relations of the more ancient prokaryotic cells enabled this event of combogenesis. Those cells had imports and exports, which led to evolution, as in the prior chapter's logic. Those new relations also set up the possibility for species of cells interacting in food webs and, specifically relevant here, for experiments of symbiosis, such as endosymbiosis. If the early eukaryotic cell was an internal recycling system between bacterial symbiont and archaeal host, that relationship could exist only between "things" that had imports and exports.

To get to the new relations of the eukaryotic cell itself, we need first to jaw a bit about a general issue that is important here and in subsequent biological events of combogenesis: *genetic conflicts of interests across scales.*[12]

Although at some point during evolution the eukaryotic cell became a relatively harmonious individual, with the symbiotic species of bacteria and archaea transformed from what they formerly were, this harmony was not necessarily the case at the start of the transition. Indeed, it could not have been the case because both types were independent and by this time well evolved to serve their own agendas. To evolve an integrated individual from narcissistic members, potential genetic conflicts at the lower nested scale of the individual members had to be tempered so that the interests that came to the forefront were dominantly at the upper nested scale of the combined cell. In this case, both bacterium and archaean had to metaphorically disappear into the larger living thing.

Indeed, the picture of the early stages of host archaean cells with their bacteria endosymbionts presents what the evolutionary biologist Neil Blackstone has called a "nightmare" of potential conflicts of interest.[13] Inside the archaean hosts lived alien, nonrelated bacteria, which could easily turn into parasites (if indeed they did not begin that way). But when the symbionts inside the host died (as cells do), they created an ongoing flood of foreign (bacterial) genes that the host would have to deal with to prevent its workings from bollixing up.

This nightmare was a kind of evolutionary awakening and birth—a transcendence. It involved the transfer of genetic control, for the most part, of the bacteria DNA into the host's central genome. The evidence for that transfer is solid. About 99 percent of the bacterial genome was eventually transferred into the host's genes, though perhaps not into a formal nucleus in the early stages of the integration. This transfer of genes from symbiont to host was key to the evolutionary event that turned the bacteria symbionts into the functioning organelles of mitochondria within the new eukaryotic cell.

From this solution came innovations to the eukaryotic cell, which, according to Neil Blackstone,[14] resulted in new potentials for its evolutionary future as a new type of individual. From the viewpoint of combogenesis, we therefore ask about new relations being gained.

SEX

Like a hard metal alloy forged from two components of softer metals, the cell union of bacterium and archaean was more potent than its parts. One result was likely true sex.

Sex is a primary innovation of the eukaryotic cell. Sex involves relations among cells of the same species. Sex is a precise kind of coordination among cells. It coordinates a mixing of genes among individuals of the same species and probably evolved as a mechanism for repair and renewal of genomes. The invention of sex could have been related to meeting the challenge of the fragmented genome that resulted from the incorporation of the bacterial genome into the union's centralized genome and eventually into what became at some point in time the nucleus of the new eukaryotic cell.

Sex seems to be have been an invention deep in the ancestry of the eukaryotes.[15] We are used to birds and bees doing sex. But it has been discovered that even single-celled, eukaryotic amoebas have sex, at least sometimes. This was long before males and females of plants or animals and apparently evolved before books would be written about the joys of sex.

Some biologists counter the claim of uniqueness for this new relation by pointing to bacteria "sex." Yes, the simpler prokaryotic cells have a number of mechanisms to exchange DNA between individual living cells. Bacteria can transfer between cells portions of their genomes via plasmids and tubes called "pili." In these mechanisms, the cells typically trade some genes (or engage in one-way giving) and then go their separate ways. But they do not have the fifty–fifty transfer between a mating pair of cells that true sex provides, nor do they have the whole-cell fusion of the eukaryotes.[16] Bacteria don't do eukaryotic sex. Eukaryotic sex is a precisely tuned dance.

NEW TRICKS FOR BETWEEN-CELL COORDINATION AND COOPERATION

Blackstone points out that success in the early evolution of the eukaryotic cell necessitated the evolution of coordination or cooperation between the prokaryotic cells that merged.[17] Like refreshed air after a storm, the eukaryotic cell could carry on in evolution in possession of new abilities.

Later in evolution, as we will see, these abilities could be applied to further mergers, this time among eukaryotic cells themselves.

As a demonstration that the newly minted eukaryotic cell was capable of further between-cell coordination, consider a second endosymbiosis that biologists also cite as remarkable. The original merger yielded the ancestor of all green algae and plants. It took a species of small, prokaryotic photosynthetic bacterium and turned its members into the cell organelles called "chloroplasts," which are in all photosynthetic eukaryotic cells.[18] It would seem that the mitochondria-wielding eukaryotic cell gained the ability to forge additional successes in coordination and cooperation between cells.

It might be worth subsuming the new relation of true sex under this general reflection on the eukaryotic cell's new abilities for coordination and cooperation. But however we parse the new relations for the eukaryotic cells, what the cells now had was a watershed of evolutionary potential, as we will see.

■ ■ ■

It's remarkable. Biologists say the evolutionary event that forged the eukaryotic cell was singular: all eukaryotic cells living today—at least those found so far, ranging from amoebas in pond scum to nerves and heart cells in our body—seem to have stemmed from a common ancestor.[19] This is odd, no? Whatever ancient conditions of Earth's marine ecosystems were able to coax into existence "our" universal eukaryotic cell might just as well have been good for coaxing other, different symbioses between cell species. So one would think. Symbiosis as a biological phenomenon would have been profligate at all times in Earth history. Perhaps the step to a level-defining, fully individual new cell type was extraordinarily tricky.[20] Or perhaps there were other, different "eukaryotic-like" experiments that went extinct. Either way, the end result, the birth of the eukaryotic cell, was for us fortunate.

I want to emphasize as a general rule as our search continues that to focus on the significance of new relations in these biological events requires *more* than asking what the pilgrim systems have that is new as they arrive on a level. In addition, we also must ask about their capabilities *to evolve* within the level: What are those pilgrims capable of evolving into? Into further lineages of species with further evolved relations within that level? Whereas all atomic nuclei could take up electrons to become atoms, not all lineages within a biological level went on to each next level of combogenesis in the realm of biological evolution.

The story about the evolution of the eukaryotic cell is still developing.[21] However these current findings carry out in the future, we do know that some sort of basic new relations or a complex suite of new relations came into existence at this level. We know those new relations were powerful because of what happened next.

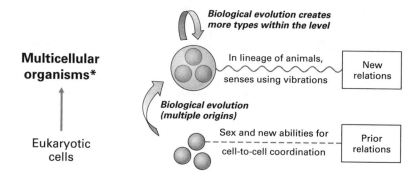

FIGURE 10.1

Complex multicellular organisms evolved from combinations of eukaryotic cells of the same species, a step taken by multiple lineages. (The asterisk refers to the focus on complex multicellular organisms.) The evolution and extinction of new lineages within the level are ongoing. Multiple primary lineages reached this level independently.

10

MULTIPLE RAMPS TO THE COMPLEX MULTICELLULAR ORGANISM

SUMMARY: Reaching this level are several lineages of eukaryotic cells, whose evolution has enabled cells of the same species to advantageously bind together, integrate, and internally differentiate. The result leads to new, larger individuals. This profound event— the evolution of a complex multicellular organism—happens multiple times. It gives the world—most prominently from the viewpoint of a woodland walker—fungi, plants, and animals. The new relations that will prove crucial to forming the subsequent level arise from the specific lineage called animals, in particular those creatures with nervous systems and senses that learn and communicate using vibrations of sound and light.

AN OPPORTUNITY FOR MANY LINEAGES OF CELLS

Things that relate might integrate. Here we will see how at this level eukaryotic cells combine and create larger organisms. These new things are not new, larger types of cells, as in the previous event, but rather integrated multicellular beings comprising many coordinated cells. *Caenorhabditis elegans*, the tiny worm that stars in genetic experiments, has about a thousand cells (known to a precise count). On a larger scale, the human body contains trillions of cells. Yet worm and human are cousins in this level.

We are in the creative dynamism of biological evolution. Again, combogenesis in biology must take into account not just what the living things have but also what they can evolve to have or become. The single-celled eukaryotes at the prior level inherited—from the prokaryotes at the origin of life—the needs and abilities to import nutrients and export wastes. The eukaryotic cells added true sex and subtle abilities to coordinate and cooperate with other cells of their same species. (I often think of sex as a subset within the new metabolic tricks.) They did not manifest the full extent of these abilities right away. Think of some of the tricks as latent, waiting for the need or opportunity. As will be seen, the opportunity occurred many times over.

Going multicellular led to living lineages thrilling to behold: plants, fungi, and animals. But out there in the wonders of nature are other, more obscure or less celebrated lineages of multicellular eukaryotes. One is the marine organism called kelp, which thrills divers exploring coastline waters, with wafting giant stalks and fronds that look like the stems and leaves of land plants. (They are even spectacular when washed up on a beach.) But kelp are genetically algae, not plants. Some species of diatoms (another type of algae, most of which are single celled) grow wonderful geometric designs of *multiple* cells.

Also, a number of lines of a group called "slime molds" have species that visually appear to be fungi but are not. These slime molds, successful as detritus feeders and often found in rotting wood and in that way like fungi, are their own unique successes into the space of possibilities of multicelled eukaryotes.

For cells of the same species to gather together and become integrated into larger, multicellular organisms was, apparently, a no-brainer for evolution. Of course, until brains were developed in animals, everything in evolution was literally a no-brainer. But this transition from cells into multicellular organisms seems to have been a cinch, perhaps because it proved so profitable.

Indeed, some biologists studying the major transitions of evolution have termed this one the "minor major transition"[1] because of the apparent ease to achieve it. From the lens of combogenesis, however, the step qualifies as fully equal to others in the grand sequence by the defining deed of uniting smaller, prior things into new, larger things with new relations.

The idea was so good, evolutionarily speaking, that even types of lowly prokaryotic bacteria took up the habit of going collective. *Myxobacteria*, likened to a bacterial "wolf pack," swarm and glide as a collective, seeking prey cells. Preparing for reproduction, multitudes of *Streptomycetes*, a common soil microbe, coordinate and convert some individuals into dead tubes to supply nutrients to others alive closer to the soil surface. Many of such collective, multicellular habits of the prokaryotes take place only during limited phases of their life cycles. But marine *Thioploca* bacteria pool huge numbers of individuals into gooey fibers, like strands in rope large enough to be visible to the eye and able to slide up and down like subway trains in the sediments to access ions as nutrients.[2]

It is not established what prevented these and other prokaryotic successes into multicellularity from getting more complex—say, with the development of tissues and organs and a life cycle like humans and insects have. Why didn't bacteria—or archaea, for that matter—evolve into things such as fish or birds? About the bacteria, the biochemist Nick Lane says, "Never did they give rise to something as large and complex as a flea."[3]

Given the core questions we have been asking iteratively at each level, the roots of what enabled the eukaryotic cells to evolve into complex multicellularity should be sought in their new relations. And what was gained in the origin of the eukaryotic cells includes what was potentially evolvable as a result of those relations. Let's consider the ability to evolve coordination and cooperation as an access ramp to what was required to get to complex multicellular living beings.

It is estimated that around twenty separate evolutionary lineages of eukaryotic cells took the "multicellular" turn in one form of lifestyle mode or another. (These lineages are themselves in most cases large genetic groupings of species.) When we focus on truly "complex multicellularity," a term the Harvard biologists Andrew Knoll and David Hewitt use, the number drops to perhaps six to eight lineages that evolved multicellularity independently.[4] These achievers of complexity are exclusively eukaryotic. Something special went on. This event clearly was *not* singular, like the origin of life or the origin of the eukaryotic cell.

And yet not all eukaryotic cells took this opportunity. Knoll and Hewitt's numbers remind us of the "glass is half-empty" viewpoint. They point out that however many eukaryotic types of cells discovered advantages to going multicellular in one or more of their specific evolutionary lines, something like sixty major groups did not do so. There are lots of "happy" amoebas out there in ponds, crawling around solo. Paramecia are successful loners, too, thank you.

So what was going on? What drove the evolution of multicellularity? And what prevented some evolutionary lineages of eukaryotic cells from going that route?

SOLVING THE CHALLENGES TO GETTING BIGGER

Cases have been made for a relatively simple logic that nudged certain lineages of ancestral pilgrim single cells to this new level independently. (From now on in this chapter, for efficiency, when I say "cells," I refer to eukaryotic cells unless otherwise specified.)

When cells reproduced by splitting, presumably for some species in certain living situations it was advantageous for the offspring cells to simply stay stuck together rather than drift apart—say, in the ocean currents.[5] Perhaps clustering aided the uptake of nutrients because some cells in a sticky collective were genetically tuned to protrude small stalks that could adhere to sediment particles, allowing the other cells in the clusters to waft profitably in the water and yet be physically secured. Perhaps the cluster lifestyle offered defense from single-celled predators because of the cluster's relatively great size. From the group might have oozed a zone of defense via waste products (perhaps inert slime). Advantages might have been only slight at first if the group lifestyle did not cost the participating cells much metabolically, if anything. If the costs of "going group" were at first free or nearly free, then evolution could drive the idea further. Sophisticated form and function might evolve.

Generalizing from such a set of advantages, the Princeton biologist John Tyler Bonner makes a profound observation: "The top of the scale is always an open ecological niche." He explains: "During the course of evolution . . . there is always room for bigger [creatures] to escape the competition with the smaller ones."[6]

But not so fast, he says. There would have been disadvantages as well. As concentrated food, collectives of cells might be prime targets for attacks. In addition, togetherness might slow down the life cycles of individual cells within the collective because the presence of others might inhibit cell division as a by-product effect. And as a group, the cells would certainly deplete local nutrients faster. Moreover, group wastes, though potentially good for defense, would also be bad as local pollution to the very cells that made it.

Another way to look at such trade-offs between help and harm considers the inherent genetic conflicts of interests across scales. We saw this issue in the evolution of the eukaryotic cell. At that event, the solution to these conflicts was related to the transfer of the bulk of the bacterial symbionts' genome into the host archaean cell. In our current case of complex multicellularity, the scales of the mandatory conciliation are between the internal cells and the entire organism. The internal cells might in a sense fight among themselves and evolve ways not to cooperate but rather to propagate their own, cellular genes.

The most accepted solution to this conflict—that is, the solution to achieve the complex multicellularity that resulted in plants and animals—is related to the fact that all the cells in the larger body ideally possess the same genome. Therefore, if some cells do not reproduce themselves into the next generation of organisms (your heart cells do not directly progress into your children), the help they give to the other body cells that do reproduce (sperm and egg) enhances their own (the heart cells') genetic selves. Genetically speaking, as genetically identical cells help others, they help themselves.[7]

There was another major challenge to success on this level. Knoll and Hewitt point out that the evolution of complex multicellularity had to tackle the problem of how to supply nutrients to cells deep inside the presumably tightly packed communal body. Those deep, interior cells no longer had direct contact with the outside to obtain fresh supplies. Depending on the species, supplies meant mineral nutrients, organic food molecules, gases such as carbon dioxide (for plants) and oxygen (for fungi and animals; but even plant cells have mitochondria that require oxygen). Cells in multicellular organisms evolved special borders to allow them to connect and perform exchanges among each other. At the scale of the entire organisms, vascular networks evolved convergently in independent lineages. We see such vasculature in the xylem and phloem tubes of plants and in the circulatory systems of animals.

Finally, organisms had to dispose of wastes. Plants release their special, net-waste gas, oxygen (good for us), through tiny pores in their leaves. We need only to look at our own bodies to see specialization for taking in nutrients and for exporting wastes in various gas, liquid, and solid forms.

By cooperating in collectives and specializing to solve the various challenges, component cells gave up certain degrees of independence. You cannot take out cells from your heart, liver, or brain and toss them as loners into a pond and say, "Go forth and multiply." They would quickly be munched up by a smarter single-celled amoeba on the prowl for weaklings or quickly die and be feedstock for even tinier, voracious bacteria still following the billion-year motto of live free or die.

Thus, the gains and losses of living solo versus living collectively have to be weighed in specific cases. We might never know what specifically drove some lineage or lineages of red algae to take one of the first turns into eukaryotic multicellularity perhaps 1.2 billion years ago.[8] We might never know what net advantage induced cells that originally were perhaps similar to today's tiny choanoflagellates to get larger and collectivize as simple animals. Those pilgrim systems about a billion years ago or less might have been similar to today's marine sponges (simple animals). Some say Earth's dismal oxygen levels prior to that time were an environmental constraint. Those are fascinating questions, but we must now consider the new relations.

WHAT WAS IT ABOUT ANIMALS?

Many lineages of cells made it to multicellularity. Perhaps six to eight of them made it to complex multicellularity. Yet using the logic in this book, only one built up from multicellularity to the next level. That one was the animal.

We know animals would come to headline the grand sequence at some point because we are animals, and this work is about how combogenesis eventually led to levels of culture. So the "new" relations we seek to encapsulate here must be those relevant to what led to us. Bear with me.

Let's first think about plants and fungi, those other successes into complex multicellularity. In some ways, plants and fungi are much like extensions of the physiology and capability of individual cells. By that I mean that the relations of plants are primarily about the passing of materials and energy in and out. Many species of plants and fungi have mastered the design trick of deploying extensive external surface areas: the tubular, fractal networks of roots of plants and hyphae of fungi, respectively; the flat solar panels of the leaves of the broad-leafed trees.

So, plant and mushroom lovers, forgive me! I don't mean to downplay the discoveries about plant sensitivities and abilities to react in complex ways. My biologist colleague Eric Brenner says that with regard to our understanding of behaviors of plants and even fungi, much of the "current paradigm is in flux."[9] Neither do I mean to ignore the amazing interconnections that plants and fungi have in ecosystems. For instance, plants and fungi help aerate and bond the world's soils. Trees affect Earth's long-term CO_2 levels.[10] Trees offer habitats for animals, such as the primates we evolved from. Yet trees did not lead—by physical embodiment into expanded degrees of nestedness—to us in the grand sequence of combogenesis.

Now consider animals. Many animals are quite well rounded as an "idea" of living form and function. OK, worms are skinny. But how animals "work"—those we might see in a wood, a bird, a deer—is different from how the extensive, external surface areas of many plants and multicellular fungi work. For many species, the animal design strategy tucks large exchange surfaces on the inside.

Consider reptiles, birds, and mammals with multiply-pleated internal organs such as lungs and digestive systems. Though this shift from outside to inside alone might seem like merely an alternative strategy to handle nutrients and wastes, the compact, three dimensionality of such internal folds facilitated the evolution of those animals in specific ways, radically enabling them to traverse the landscape as whole things. Here we focus on animals with nervous systems and freedom to move around a lot.

The animal's nervous system: a high-tech, magical organ, indeed. A nervous system allows some major new relationships. In fully mobile animals, the nervous system facilitates movement of the entire creature through the environment. Many animals crisscross territories. They travel trails. They navigate spaces, in many cases rapidly and deftly. A dolphin leaps. A dragonfly reverses course in a midair blink. A lion pounces.

Relevant data come from the life cycle of the small marine animal called the tunicate or sea squirt.[11] When a larva, it has a simple nervous system and briefly swims. When maturing, the larva attaches itself to something—say, a suitable rock. From a swimmer in youth, it becomes a filter-feeding, sedentary adult. The shocker is that the creature now self-digests most of its nervous system because that system is not needed and is costly to maintain.

The ancestors of mobile animals might have been something like today's marine sponges. Sponges, too, are sedentary. They lack nervous systems. Couch potatoes, beware. Underwater sponges are marvelous to witness when you are snorkeling above a reef. But compare the thrills of a crab, puffer fish, moray eel, or octopus—well, we see these animals, but they also often see us. And many more see us than we see them. This "seeing" is a new kind of relationship to a thing's surroundings.

For the next level, a nervous system is crucial not just for movement but for learning (I include here learning by crabs, fish, octopi). But it's worth highlighting the certain organs of relationship that many complex animals have—including our ancestors—and that are integral with the function of movement (or learning).

Consider: movement itself is not particularly special. For example, the single-celled eukaryote amoebas also move as whole bodies, as do paramecia and even many single-celled prokaryotes. Bacteria such as *Escherichia coli* spin to randomize their direction and then can sense dark and light and nutrient gradients and so move accordingly. Yet none of these single-celled creatures has a nervous system. To find something truly special in "higher" animals (those that will get to a subsequent level on the grand sequence), we might look at the ability to look.

Many animals have abilities for precision remote sensing. Yes, there is direct touch, so important in sex, for example. But think about sight via light, hearing via sound waves, and smell via molecules wafting through the air. Such senses allow information about specific other organisms' bodies (or their effects) when those others are distant from the sensing creature's body. Signals across space become possible with such senses. Smell is particularly crucial to ants (chemical trails), to dogs and bears (and most mammals), and even to some birds (for example, vultures). In our own primate lineage, smell downgraded a bit in evolution as eyesight gained special prominence. And for what is to come in our unfolding of the grand sequence sound and sight were special.

Eyes and ears. Those organs are highly specialized, like the internal lungs and heart, and mediate direct relations between creature and its environment. To emphasize, those "remote-sensing" senses are organs of relationship that do not require physical contact of bodies. There is something extraordinary, it will turn out, in those senses that allow distant connections to others and environmental patterns across space. In the transition from biology to culture, sound and light will eventually serve as the media for fully sophisticated language. This is possible because light and sound travel through air and water as vibrations.

■ ■ ■

It is fascinating that current understanding says that all animals are monophyletic.[12] They all evolved from a single ancestral lineage—one out of those perhaps six to eight separate times that cells reached the various types of complex multicellularity we have today. I wonder why. Why was there only a single beginning to that basic design "idea" or "concept" of the original animal? In contrast, both land plants and certain multicellular algae, such as the marine kelp, independently evolved the physical design "concept" of fronds and leaves as a style of photosynthetic multicellularity.

Leaving to the future this part of the many mysteries of animals, for our purposes we see that in the animal as design idea evolution eventually put in place organs to use vibrations of sound and light. These relations were new and crucial. The organs in question are the eyes and ears, extensions of the nervous system at the core of this innovation, making and interpreting the vibrations, coordinating messages and knowledge, creating and planning action, and learning. Today in the sensory worlds of animals, we hear thumpings, we see displays of color and dance. There are signals of danger. There is honking to keep in touch with traveling comrades during a long migratory flight. These relations set the stage for what is to come next.

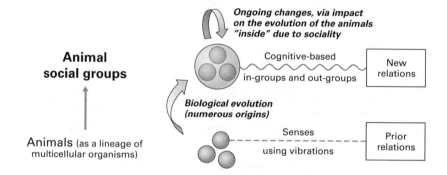

FIGURE 11.1

Animal social groups consist of animals (one "special" lineage from the prior level of complex multicellular organisms). Biological evolution is operating, but its workings here are through the evolution of animals in the context of being social (the ability for "social learning," for example). The "cognitive" new relations refer to certain lineages, which, in addition to many others, certainly include those that lead to human ancestry.

11

ANIMAL SOCIAL GROUPS WILD WITH POSSIBILITIES

SUMMARY: Following fast on the evolutionary heels of the origin of animals are animal social groups. Within the things of this new level, animals use senses to know the whereabouts of friends, foes, and foods across distance and advantageously bond into networks of individual animals of the same species. These groups or societies mold the evolution of animals within them as the animals gain capacities for social learning vital for their individual success. The new relations that are crucial on the social level focus on the creation of in-groups and out-groups by shared behaviors and correspondences in the recognition of membership.

A RAINBOW PARTY OF TYPES

Prokaryotic cells merged into the eukaryotic cell. And then eukaryotic cells merged into the complex multicellular organism. Each of these events was powered and enabled within evolution by styles of "mutual aid" offered by going collective and integrating. Next, following the logic, we focus on animals. Animals merge into new things.

What are these new things? There's a wonderful spectrum of types. They range from simple herds of antelopes and flocks of birds to complex social insect colonies, with their division of labor and high integration and interdependence among the individuals. Despite this variety from loose associations to societies, almost like animal bodies that vary from simple to complex, here I conveniently call them all "social groups."[1] They are groups that contain social interactions.

Ocean life crosses the entire spectrum. Yes, perhaps *social group* is too strong a term for a coral colony, but the individuals of the same species do secrete a group "creation" of hard calcium carbonate sculptures upon which the tiny animals live as a soft outer carpet of delicate, water-filtering polyps. The gorgeous sea fans—a type of coral that waft in the underwater currents—are also colonial animals. These swaying fans, some a meter across, give hundreds to thousands of small individuals the communal ability to collect nutrients from the water, similar to the strategy of how green terrestrial leaves collect CO_2 and sunlight. In addition, fish school. Collective living, as an option, can tender advantages to predators and prey. Tuna collectively hunt and can strategically herd prey. Anchovies can school for defense to confuse their predators.

Colonial animals such as coral heads and fans and even schools of fish are relatively simple societies. At the other end of the ocean's spectrum are the strange siphonophores and the eusocial shrimps. The siphonophores are to the individual animal what the multicellular animal is to the individual cell. To the uninitiated eye, a siphonophore looks like a single, floating, tubular animal, but in detail it is a population of clones of many small animals who during development stay bonded and then differentiate into functions such as locomotion and feeding. Among the eusocial shrimps, a "queen" lays most of the eggs. The other shrimps help raise the young. So there is division of labor and the obligation to live together, unlike, say, a fish that can at least go away from its school and chill out by itself.

The word *eusocial* is a twentieth-century coining, meaning "good-social" (we saw the Greek prefix *eu-* before in the term *eukaryotic*). A eusocial animal group has such a high level of integration that it is sometimes called a "superorganism," especially in reference to the terrestrial eusocial insects.[2] For example, in some species of ants, their queens lay all or most of the eggs. Various functional castes often include female workers and guards, and males are just for mating. The colony is a superbody made of many individual bodies.

Societies on land are abundant at the simpler end of the spectrum, too. Herds of wildebeests, mobs of kangaroos, and flocks of birds—all of these groups are like the schools of fish in the sea. When the same patterns are invented multiple times by evolution, the phenomenon is called "convergent evolution." For example, large herbivores have evolved everywhere on Earth, with similar social behaviors for migration and protection. Thus, we find grazers that group, such as Australia's kangaroos, North America's bison, and Africa's antelopes.

Eusocial animals with their reproductive specialists and other castes are also evolutionary convergences: the eusocial marine shrimps; the terrestrial insects such as certain bees, ants, and termites; and even naked mole rats, mammals that live underground with their queen and feed on tubers that they reach via their communal tunnels.

The colors of social spectrum—from loose to highly integrated—have evolved multiple times, showing the wide utility of particular patterns. For example, communal, cooperative breeding is thought to have arisen independently nearly twenty times in clades of spiders and even more times in birds.[3] Full-on eusociality has arisen independently nineteen to twenty times when we include all instances in termites, ants, bees, wasps, aphids, ambrosia beetles, shrimps, and the mammalian naked mole rats (and most of those "instances" include multiple species derived from a common eusocial ancestor species).[4]

For animal evolution to give rise to social groups was apparently very easy. The social group as a thing, system, entity, ontum was multiply invented by numerous evolutionary lineages of animals. We also saw this pattern of multiple invention previously at the level of multicellularity. For animal social groups, the independent achievement of the new level is even more numerous. And as we saw with multicellularity, for social groups there is a range from simple to complex. Animals have been able to explore a rich "pattern space" of possibilities in the evolution of their social structures at this level.

RHYTHMS TO THE CALLS TO COMMUNITY

One constant across the levels of life is the basic need to satisfy the breakout new relations that originated in the simplest first cells: the imports of nutrients and exports of wastes. If animals of the same species can help each other achieve these basic physiological needs—by any means, such as enhanced protection—as well as the need to reproduce, they might go collective.

The building of the first communities of animals probably happened very early in animal evolution. For example, the fossil records from hundreds of millions of years ago show the strange marine trilobites, which look a little like today's horseshoe crabs, heads to rears in a group parade, like boxcars in a train, migrating presumably.[5] Once animals could coordinate with their environments using senses, they might find advantages to coordinating with others of their species. They would gather into communities. Ancient trilobites (all now extinct), for example, had complex eyes on stalks. Such animals also likely gathered in groups for mating. The group pattern must have offered diverse opportunities. Along those lines, perhaps mating itself might be *the* primordial animal social group, from two at minimum to what in some cases are orgies.

There are many potential advantages to group living: warnings to hide (prairie dogs), aggressive defense (fire ants), collective hunting (wolves), information gathering and exchange (bees), group architecture (coral, termite mounds), support in migrations (wildebeests, birds), gatherings for matings (bird leks).

Much at this level sounds strikingly familiar to the previous event of combogenesis that took cells to complex multicellularity: feeding, breeding, moving, defending. For evolution to shape animals into social beings, there must be advantages to sociality for the individuals' reproductive potential as related to specific networks of genes possessed by those individuals (which can be common among individuals). As a bottom line, it has to be worthwhile on evolution's balance sheets for the individuals to participate in the higher-level social system.

This point hearkens back to the insight by John Tyler Bonner that "the top of the scale is always an open ecological niche."[6] Getting larger might mean evolving larger bodies within a level (simply a larger animal, say). Or getting larger might mean evolving a behavioral repertoire to join with others to form a larger "body" that happens to be a social one.

The balance sheet holds disadvantages, too, though. A high density of individuals can deplete local resources. Diseases can run rampant. Wastes can reach toxic levels (that's why humans invented sewage treatment).

The group is a billboard for predators seeking a place to fill up their bellies. And there are issues for how the group will police within itself against internal "cheaters" or "free riders," those individuals who try to reap benefits of the group without contributing to it or who simply grab so much they harm others. Generally speaking, this list of detriments is a replay of a similar list we saw for cells going multicellular.

In the evolution of the multicellular organism and now of the animal social group, advantages and disadvantages are weighed out and then juggled in various ways. Many cell lineages stayed with the single-cell form as the successful mode. In many species of animals, individuals spend their lives basically alone except for mating. But then many lineages of cells went collective, as did many lineages of animals.

Evolutionary biologists have been fascinated by such commonalities across what they consider "major transitions in evolution." Biologists Andrew Bourke and David Sloan Wilson have separately emphasized that these transitions are in essence social innovations at various scales in which a general pattern repeats.[7] Yes, patterns remarkably repeat. Yet the differences are what make the grand sequence a sequence. To those differences we now turn.

THE APPARENT SIMPLICITY OF ANIMAL SOCIETIES

One feature of animal societies stands out in contrast to multicellular organisms of the previous level. Even the most complex and sophisticated animal groups seem rather simple when considered by the number of internal, functional types. Eusocial insects—for example, a honeybee hive—operate with merely four, five, or six functional castes, according to current knowledge. A wolf pack has alpha, beta, and perhaps gamma males and females, juveniles, and young, according to wildlife biologists. Compare those numbers to the types of cells in trees or mammals, which have hundreds or more functional types of cells in their bodies.[8] Merely in our brains, you and I might have a hundred or more types of neuron cells.

Furthermore, cells in the multicellular organism are arranged in nested groups called "tissues," and those groups might be in organs with multiple parts (the heart with its muscles and valves). So even though the prior step from eukaryotic cell to complex multicellular organism must have started in a simple way, with sibling cells simply sticking together, many multicellular beings eventually evolved to be so internally sophisticated it bends the mind. One result is our complex bodies, which often present great puzzles to doctors. In contrast, animal societies seem rather simple, as sophisticated as an ant hill is in its own special way.[9]

I see several potential reasons for the relative simplicity of animal social groups. (Again, this metric considers the number of differentiated "parts" one might count in the group, obviously not the total numbers of individuals, which can be enormous.)

First, in the social groups ultimately of interest here because of similarity to us, the individuals remain as individuals who can feed and move around. This maintenance of individuality is related to the fact that they tend to keep and use their senses to navigate their environments.

Consider individual ants with their hard exoskeletons. It would be very difficult for evolutionary forces to select for physical merger of the ants of an ant colony. And much of the colony's skill as a whole requires that scouts forage more or less on their own and then share information about where they find food. Then they all participate in laying down their famous roadways of odors to alert the colony to a food source. Effective group dynamics is keyed to a degree of freedom for each ant. In any case, their insect exoskeletons would make merging difficult. The same point can be made for most land animals in social groups: it is hard to imagine how elephants, wolves, or eagles could physically merge into a single body.

It is interesting that physical merging was more evolutionarily easy for marine creatures. Water moistens surfaces and can supply nutrients by simple diffusion or currents. In coral heads and sea fans, animal members physically touch with some degree of interconnection. As noted, an extreme example is the siphonophore, a social body with individual animals physically bound but functionally differentiated, such as the species the Portuguese man o'war.

Another reason for social simplicity is that member animals are often genetically distinct individuals. And most members are able to reproduce, with the exception of eusocial animals, which have "queens" that do all the reproducing. For animals in so many cases, remaining genetically distinct individuals makes the genetic conflicts of interests across scales more difficult to overcome than at the prior levels.

We saw at the prior levels the repeated theme of genetic conflicts of interests across scales being solved in various ways. Individuals may benefit from going collective (as happened when some prokaryotes went eukaryotic, some eukaryotic cells went multicellular, and some animal multicellulars went social). But if those individuals are weakened in the metric of genetic gain or loss by going collective, they might remain solo, as many lineages have done at all those scales. Solutions that led to prior levels depended on the specific challenges. Recall that in the evolution of the eukaryotic cell, the genome of the bacterial symbiont got transferred into the central DNA of the archaean host cell. In the evolution of complex multicellularity, the cells of the multicellular organism have the same DNA, so by helping each other, they are in a deep, genetic sense helping themselves.

In general, individuals animals in social groups are more closely related than any random batch drawn from the entire species. Yet because of the differences among individuals in those social groups, they usually compete with each other as registered on the genetic balance sheets for future generations (exceptions include populations of aphid clones and many of the clonal marine colonials). Genetic variation among individuals in most social groups helps genetically distinct cheaters or free-riders to thrive. The balance between advantages and disadvantages can easily tip one way or the other. Small alterations in the genetics can often alter the balance depending on ecological circumstances.[10]

Andrew Bourke pointed out to me still another reason for the relative simplicity of animal social groups. It has to do with total numbers of individuals in the collective. The number of cells in multicellular organisms can reach many trillions (30 trillion in the human body, as noted in chapter 1). Individual organisms in an animal social group rarely get beyond 10 million. Thus, Bourke argues for a "causal relationship between size (number of subunits) and social complexity." Then differences in numbers, like those just cited, "alone might help explain the relative simplicity of animal societies."[11]

One final reason for the apparent simplicity of animal social groups: groups usually do not die as a group or single unit. Some, but not all, eusocial insects are exceptions here, with actual life cycles.[12] In a sense, the group in many cases can potentially be immortal. An example is an anthill that I have known and even befriended (at least in my mind) for decades and that was "alive" for decades before that.

When the group is not inherently mortal, the situation is weaker for the powerful scythes of natural selection to work their pattern-shaping marvels. The issue requires more study, but it seems to reasonable to posit a relationship between being subject to a whole-system ax of death (predators that can eat you in one quick gulp or fell swoop or a life cycle with senescence) and the evolution of internal complexity.

All these angles likely interact to degrees. As one result, the chimpanzee community seems not as complex as the chimp's own physiological miracle of body with lungs, heart, and brain. Society as a system of individuals seems more fragile than the body as a system of cells. But as we move toward human culture, we will see that in the "weakness" of the animal social group lies its strength in terms of generativity to the next level.

THE NEW RELATIONS: COGNITION BASED, COLLECTIVE IN AND OUT

Though I emphasized that most animal social groups seem rather simple, with nothing like the human body's complex organs, there is a flip side. The individual animals, as noted, can be extraordinarily complex. So can their behaviors. That's because one of the organs of those bodies is the brain.

Behavior as tied to the brain can certainly be sophisticated in animals that live mostly solo. Octopi are excellent examples. For animals dependent on group living, that sophistication often involves social learning. Social learning is a special kind of learning that takes place only by growing up and getting along in the group. Most birds and mammals have crucial social learning in the preparation for "being on your own" that goes on at least between mothers and offspring (many bears) or between both parents and offspring (many birds). In highly social animals that spend most of their lives in groups, this learning can be never ending and involves larger numbers of others.

Social learning is a kind of attunement. Animals with such lifestyles need the group to learn what to eat, how to hunt, how to get along in hierarchies, how to entice success with potential mates, and more. Critters watch each other, observe patterns, build trust, learn wariness, test personal boundaries, form and use memories of gold medals or agonies of defeat in dominance trials. Thus, in seeking what essential new relations emerged on this level, we might inquire into the shared psychological patterns of certain animal groups' membership. After all, we would be looking to designate relations of the group as a whole.

Consider animals that possess the following characteristics:

- They are together all or most of their lives.
- They have specific, diverse, and variable relationships with other members of the group, derived from actual experiences with those others, and they can even be affected by relationships that others have among themselves.
- There is some sort of collective control over who is "in" and who is "out." In these groups, a stranger does not just walk right in and sit down.

An amazing video by *National Geographic* shows the first year of a wolf pack in Yellowstone Park after the gray wolf was reintroduced there in 1995.[13] The pack loses its alpha male to a human gun outside the park boundary. A foreign male from another pack seeks to join. His life hangs on a thread. He is tested by the others, one by one, until he is finally accepted. The moment can still bring me to tears. The pack as a group let someone new in. It could have said "no" and proceeded to kill.

What we seem to have in such interplay is a collective based on shared psychological patterns of who is in and who is out. The result is a social boundary. It's not physical like a wall, but it's still highly functional. Such a boundary preceded the more complex in-groups and out-groups of human societies, which serve as endless data for sociologists because of our labyrinthine social capers. But I suggest that with animal groups such as the wolf pack, we surely see the start of a pattern of "we."[14]

I am not saying there is a conscious "we" in most animal social groups. Good heavens, no! But a "we" exists pragmatically in the common patterns among the individuals who are able judge each individual as a "yes" or a "no." In a sense, those judgments are being renewed and mutually reinforced every day that a group such as a wolf pack stays stable in membership. We can witness in behaviors the presence of cognitive patterns or maps in the individuals. When the maps correspond (at least roughly) across individuals, then the group can stay stable in structure. There is then shared recognition of membership.

And of course things change, as the wolf pack in the video demonstrates. All is not continuous harmony in such groups. In a pride of lions, an outsider male lion (or a small group, perhaps brothers) can take over the pride, and all hell breaks loose, including the murder of cubs. The point is that shared psychological mapping of "yes" and "no" by way of who is admitted and who is excluded creates a group whose structure is so clearly delineated that primatologists can arrive at a jungle camp, observe, and say, "That's a chimp community. It has this territory; it has enemies in other communities," and so forth. The group is a system of individuals, together, based on psychology related to the ongoing adjustments enabled by the capabilities for individuals to engage in social learning.

■ ■ ■

The general category of social structure into which some anthropologists place chimpanzees is a fission–fusion society.[15] It is special for us. It is surely relevant to our own social evolutionary roots that our closest living relatives, the chimps and bonobos, have fission–fusion societies.

Animals in a fission–fusion social world can enter and leave shifting and flexible subgroups within their overall larger group. In chimpanzee communities, for example, groups called "parties" can spin off as smaller units for days. Yet when the party reunites with members of their larger community, there will be greetings between them. These smaller parties can quickly change in membership and size. The negotiation of relationship, entangled by past interpersonal events, is therefore ongoing.

Other examples of animals that possess fission–fusion social structures are spider monkeys, spotted hyenas, elephants, dolphins, and bats. Note that this list includes animals that, at least in some cases, are known for their keen intelligence. My discussion of shared psychological maps of social boundaries connected to social learning, however, would not be limited to the formal fission–fusion social animals. I obviously would include wolves and many, many other species. But our human heritage is shared with apes in fission–fusion systems that have particularly intricate, flexible, and negotiable in-group and out-group regulations and internal adaptabilities to form and dissolve smaller groups at various scales.

Though the animal society seems to be a weaker level in terms of its internal complexity, that weakness is its power. Cognitively based "borders" of social in-groups and out-groups might seem loose and indefinite compared to the touchable skins of animals' bodies, the secure membranes of eukaryotic cells, and the firm walls of bacteria. Yet we are at the verge of the next level. We are at the verge of what I later call a new "dynamical realm," equal to the jump from molecules to life. From this seemingly small step of animals into societies will come a giant leap of combogenesis.

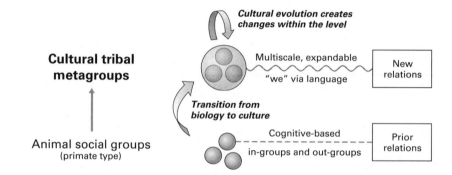

FIGURE 12.1

The cultural primary metagoup starts in its primary form when the word *we* is used for culturally connected groups larger than the daily group. Transitional dynamics occur as biological evolution transitions into cultural evolution.

12

TRIBAL METAGROUPS AND CULTURAL EVOLUTION

SUMMARY: By the logic of the grand sequence, this next transition would build outward. Thus, from animal groups of the prior level, we would expect groups of groups, perhaps webs of groups. Proposed here are the geographically extended networks of early human societies as the next level: tribal metagroups. The key to this level is the ability to belong to, at minimum, two scales of groups, enabled by the conscious, expandable "we" that can go beyond the daily group. Language is instrumental in the innovation. So are cultural artifacts, both material and symbolic. The result is the birth of cumulative cultural evolution.

FOLLOWING THE LOGIC, BUT TO WHERE?

If we continue the logic of the grand sequence, what would we expect next? This next level, like those before, would somehow encompass the previous levels. Thus, animal social groups would now be components of the new level's things.

During combogenesis, prior things change when they merge into larger things. The prokaryotic cells that joined in endosymbiosis to the eukaryotic cell radically evolved. The eukaryotic cells evolved into quite new types (nerve cells, for one) within the complex multicellular animal. Animals that live in social groups evolved to be tuned to that mode of living. Research has shown, for example, a positive correlation between primates' brain sizes and the sizes of their social groups, a finding that has led to a social brain hypothesis—namely, that larger group size was important in driving the larger human brain in its early evolution in our hominin lineage.[1]

So we expect the animal group itself to change when it merges into something bigger. But what is a larger system composed of societies? It is not immediately clear how a grouping that contained animal groups could yield anything substantially new. What could a society of societies create that is novel enough to be worthy of a new level?

Most scholars involved with the evolution of human culture recognize social innovations as hugely important. I suggest that the human way of creating and living within larger groups made of smaller, daily groups qualifies as this next level of grand sequence. This change is not just a matter of being "novel enough" to be called a new level. It is so revolutionary that a new type of evolution begins: cultural evolution. Later in the book, I call this level the base level of a new dynamical realm.

Scholars, such as anthropologists, employ various terms for a new social modality that is and was special to humans. Data are gathered from humans in living hunter-gatherer groups today, from the recent past, and from paleoanthropology. One overall aim is to understand our deep heritage. Here is a list of terms for what is socially special:

- *metagroup*[2]
- *hypersociality,*[3] *ultrasocial*[4]
- *megabands,*[5] *multibands, metabands*[6]
- *tribes* as bands of bands[7]
- *clans*[8]
- *type II superbrain*[9]
- *we-intentionality*[10]

This is not the place to parse those terms across cases. The words do show that many scholars consider that an innovation in scale is relevant to a primal structure of human sociality: *meta-, mega-, hyper-, ultra-, super-, multi-*. In choosing a term for this new level, I decided to go with *tribal metagroup*. The anthropologist Kim Hill essentially uses the terms *metagroup, metaband,* and *tribe* interchangeably.[11] *Tribe* is a familiar word. And I like *group* because it can apply to subsequent levels of combogenesis and not be limited to the "band" of hunter-gatherers. The prefix *meta-* indicates the multiple, flexible scales of this "thing" and a scale beyond the smallest daily group.

To be a new level, the tribal metagroup is dual scale at minimum, as I discuss. Once in place in the human story, the tribal metagroup developed or evolved out of its earliest presumed structure into more intricately sophisticated structures of the ancient and even modern world.

THE EXTENDED WEB OF CULTURE

Let's peer back in time to some facts generally agreed upon for that propitious, extended dawn of early human cultural evolution. During the time of the Upper Paleolithic, say 40,000 years ago, humans had been fully anatomically "modern" for more than 100,000 years. (Though we cannot be sure of no major genetic changes during that interval that might have given rise to crucially new cognitive capabilities; this issue is debated.) But if the skeletons of people did not change by much during that time, the artifacts certainly did.

In the fossil record of the Upper Paleolithic, paleontologists find burials. They find finely worked bone tools for daily activities, such as needles for sewing and fishing hooks. They find a varied tool kit of small, stone "microlith" tools specialized for the numerous tasks required by complex lives of ancient hunter-gatherers. Even earlier, in the African Middle Stone Age, there is evidence for snares or traps, bone tools of awls and projectile points, and eggshell and ochre fragments with geometric designs. The archaeologist and human paleoecologist John Hoffecker concludes, "In sum, archaeological evidence for fully modern cognitive faculties (with the exception of representational art) is associated with anatomically modern people and the African [Middle Stone Age] between 75,000 and 60,000 years ago."[12]

During those distant times, humans lived in small bands. Those bands likely ranged from about fifteen individuals in a single foraging group to perhaps fifty in an "overnight camp group."[13] That would put the scale of the basic human band roughly in line with today's wild chimpanzee communities. So humanity was not a big win in that count.

But then consider what must have been going on from the angle of the artifacts. As one example, consider the Upper Paleolithic's palm-size sculptures of females called or miscalled "Venus figurines," found across Europe and parts of Asia and spanning an interval of perhaps 20,000 years. They might cause a Picasso or Brancusi to swoon. Whatever these enigmas meant, they show us that symbolic traditions were spread across wide geographical areas over a staggering interval. The figurines that have come to us are not cookie-cutter identical. Yet quite a few (although not all) share obvious characteristics such as their facelessness and voluptuousness.

It is difficult to ponder them and not see something going on that was larger than the local daily band. More than biological mates were being exchanged among these culture-possessing populations.

Even earlier, in the Middle Stone Age (or Middle Paleolithic), about 70,000 years ago, people in southern Africa crafted beads of perforated mollusk shells. Similar artifacts, made of the same species, have been found thousands of miles away in Morocco.[14] Paleontologists have compared these artifacts and what they say about trade to modern beads of ostrich eggshell in the wide-ranging circles of trade and gifts by hunter-gatherers in today's Kalahari Desert.[15] Such distant links beyond the confines of small local bands might have been in place for a very long time indeed.

Randall White, an anthropologist at New York University and expert on the early Upper Paleolithic, has told me he imagines a scene involving a traveler-stranger approaching "your" band.[16] The stranger is not totally alien, though, because he or she wears a choker with fox teeth prepared and strung like the chokers your own group makes. Your customs extend way beyond your valley. Thus, greetings and cautious acceptance follow— unless you have heard in advance that a "cheater" journeyer was coming close. We do not know if that larger, extended metagroup, that ancient tribe, would have a name. But it wouldn't be surprising. At some point, such naming did happen.

The profuse cultural web would not have had political meaning in any modern sense. The network would have been diffuse because of low population densities. Nonetheless, these cultural traditions that linked people across regions were embodied in tool kits and manufacturing techniques as well as in symbolic artifacts and eventually cave art. Such traditions show properties of a system, a thing, that go beyond the dynamics of local biological kinship and spatially beyond the local daily survival group.

Studies of contemporary hunter-gatherers by anthropologists have revealed the import of enforced egalitarianism as a common rule. The anthropologist Christopher Boehm deems that egalitarianism emerged around the sharing of big-game meat.[17] Once started, the ethos of sharing can spread far out from the local daily group to a tribal metagroup. Eskimos (Inuit) are members of survival networks across thousands of miles. The archaeologists Kent Flannery and Joyce Marcus contrast the extended networks of sharing in human hunter-gatherers with food sharing in chimpanzee troops: "It [food sharing] does not, however, extend beyond the limits of the troop. No one has ever seen members of two chimpanzee troops meet at the border between their territories and exchange food."[18]

Recent work by the Arizona State anthropologist Kim Hill and his colleagues has revealed crucial aspects of tribal metagroups. They studied current hunter-gatherers: the Ache in eastern Paraguay, South America, and the Hadza in Tanzania, Africa.[19] Average band sizes are twenty to thirty-five individuals, living about ten to thirty-five kilometers apart. Yet over a lifetime an individual will have a "social universe" consisting of likely more than a thousand individuals because of the huge network of interactions within the metagroup. During their lives, the researchers found, human males will be able to witness tool making typically by three to four hundred other males across the tribe. In what to me is a dramatic contrast, as it was to the researchers, typical chimpanzee males interact with only about twenty other males in their entire lives. This expansion of interaction is the potential and operating power of the metagroup, a multiband system, a group of groups, across geographical space and over a long lifespan.

LANGUAGE AND MATERIAL THINGS

Language is one huge eruption of cross-geographical unification in the full launch to this cultural level. Virtually all analysts of the transition into culture focus on language as a factor.[20] So does the Bible: "In the beginning was the Word." The word in that famous sentence was actually *Logos*. Language is tied into the logic of the dynamics of this new level.

Language plays an irreplaceable role both within and across groups at all scales and thus is vital to the social metabolism of metagroups. Language acts somewhat like the way that the color force of quarks and gluons binds them into nucleons and then a residual color force binds those nucleons into the still larger atomic nuclei. Language can bind locally with full strength and then with lesser strengths forge links across distance.

The innovation of signals that evolved in animals that were our ancestors facilitated new relations that rode on light and sound within those animal groups. Human spoken language emplaced ultrafine modulations upon basic sound waves. Light carried signals from gestures. Grunts and calls and perhaps sung notes became words. With increasing distance and geographical barriers, such as perilous rivers or mountain ranges, language groups became unintelligible to each other. But certainly across a landscape, language helped build and maintain the new, large sizes we are interested in on this level. Indeed, according to the British researchers Robin Dunbar, Clive Gamble, and John Gowlett, the ancestral language "tribe" was likely about 1,500 individuals.[21]

A lively debate is ongoing about the beginnings of language.[22] That is not our concern. Here, within the new level, we note only that full language was likely needed for the metagroup by, say, the Upper Paleolithic or, as we have seen from the evidence, even earlier in the Middle Paleolithic (Middle Stone Age). The archaeologist Lyn Wadley confidently envisions Middle Stone Age people speaking to each other at their habitation sites in southern Africa.[23]

So much that language does! It is vital for the propagation and improvement of tools and symbols. It allows complex scenarios about daily lives to be shared. People talk about the past. People can jabber, pro and con, about others not present that day or even about a relative when news arrives, though the relative has not been seen in years. Through memories, ancestors stay alive. People can name thousands of things in their natural and crafted universes. They can discuss goals involving those things and each other.

The objects of cultural space and time are entangled in expansive networks involving the mind and thus are well beyond the networks of immediate sensations registered by eyes or ears.[24] It is easy to see how language, using its ultrafine modulation of sound, as well as visual arts and crafts that modulate vibrations of light, are progressions on capabilities from the prior animal senses. The senses get extended by what psychologists call "mental time travel."[25] Past, future, and hypotheticals take on membership within these new intricate lives.

In this expanded and expandable "we," material artifacts as well as language were crucial. The archaeologists Fiona Coward and Clive Gamble have argued that to evolve sophisticated societies from a "shared pattern of face-to-face social interaction," a "key process was the gradual incorporation of material culture into social networks over the course of hominin evolution."[26] In the Harvard psychologist Steven Pinker's discussion of the human "cognitive niche," language, tools, and "hypersociality" operated together in a way powerfully special to humans.[27] Language and the system of increasingly complex artifacts cofacilitated each other and presumably helped link the small, prior animal societies into the extended tribal metagroups.

CUMULATIVE CULTURE AND CULTURAL EVOLUTION

Kim Hill also says, "High intergroup interaction rates in ancestral humans may have promoted the evolution of cumulative culture."[28]

The term *cumulative culture* is used to distinguish human culture from the simple cultures of some animals, such as the intriguing chimpanzee use of sticks to fish out yummy termites from inside their mounds. Cumulative culture improves and is seen in the advances from human ancestry, such as the progression from simple hand axes to more refined tools, including microlith kits.

The metagroup is important in cumulative culture to keep those jabbering tool makers in contact with enough other talkers. Recall the three hundred tool makers an average crafter would witness in the Ache and Hadza. If groups are too small, traditions die out. People need to be in contact with a large enough population that they might see innovations or at least be able to learn, copy, and pass along best practices. Hill emphasizes the causal connection between the expanded size of the metagroup and cumulative culture: "larger numbers of cooperative interaction partners [do] drive cumulative culture."[29]

For this larger contact, there had to at least two scales to the human metagroup.

First, the local band. This band stayed small, in line with what was probably our primate ancestral band size, because of limitations in food and local resources.

Second, the larger group. This group was required for the maintenance of language, tool making, ritual relationships, occasional large gatherings of the metagroup—in other words, cumulative culture in all aspects from material to mental and social. In its primary incarnation in human ancestry, this larger group would have been geographically fuzzy and fluid in its extent.

The concept of cumulative culture implies cultural evolution. We saw how the new relations of prokaryotic cells virtually implied biological evolution. A fairly sophisticated culture that requires a metagroup social structure virtually implies cultural evolution. Individuals vary and will vary their productions—tools, sentences, symbolic crafts, details of relationships with others. There will also be selection. People make decisions. Everyone cannot copy everything from everybody. Language allows yes and no to occur, both by the individual and in collective decision making. Thus, there is the propagation of patterns, the variation of patterns, and the selection of patterns.

Propagation, variation, and selection are the three components to a general evolutionary process, which we will look into more in part 3 of the book. That general process has both biological and, with the tribal metagroup, cultural forms. The metagroup's scale, cumulative culture, and cultural evolution are interconnected. Hill and his colleagues say, "The emergence of metagroup social structure might explain why humans, but not other social-learning animals, evolved the cognitive mechanisms that produce cumulative culture, and why *Homo sapiens* were able to replace other hominins as they spread out of Africa."[30]

THE UNBOUND "WE": NEW LEVEL AND NEW RELATIONS

These early cultural systems do not have well-delineated or edged borders. Thus, the resulting new thing, system, entity, ontum on this level is tricky to hold in mind as a thing. It is not physically bounded like the other just-prior things on the grand sequence, such as the prokaryotic cell, the eukaryotic cell, and the multicellular animal. (Maybe the metagroup fades indefinitely outward like the electrons of atoms; but, of course, unlike the electron, it cannot be theoretically anywhere in space–time, at least not yet.)

There is a breaking away from "boundedness." Such a breakout from the definite borders of cells and animal bodies was previewed in the animal society. But the tribal metagroup is an ontological "creature" even fuzzier and different from the famously fuzzy dispersions of ants in colonies (to take an apt example from the prior level of animal groups). Moreover, the multiple scales of the human "we" at the level of the metagroup exist nested within each other with varying degrees of fuzziness. The component human bands can move around and shift membership in coarse and flexible territories; the overall cultural tribe scale is quite fuzzy. This fuzziness, however, directly relates to the new relations that come in with the metagroup.

But, first, how does the metagroup work in the model of combogenesis and the grand sequence we are formulating here?

The grand sequence is defined by a series of ever-larger, new systems built from prior things.

For the origin of human complex culture, the animal social group had to be there first (again, in the Middle or Upper Paleolithic; I am ignoring this transition's entire ramp across time—for example, those fascinating issues about tools and *Homo erectus*[31]). Animals that were our ancestors related in a network of social learning, had a complex fission–fusion structure, and used all the senses, but particularly relevant for the emergence of the metagroup were the media of light and sound for the ongoing cognitive entanglement with others possessing unique personal histories.

In those primary human metagroups, bands approximately in line with (or smaller than) the sizes of chimpanzee communities became part of what eventually was a much larger, tribal matrix. But the existence of culture did rely on those bands as basic units for the ultrasocial humans. The overall metagroup of a thousand people might have contained a dozen or more bands. Those thousand people were not just massed together like a crowd at a rock festival. The animal social group thus carried on as a component structure up into this level, as a type of node within the larger network. Of course, the animal group changed, as all units change when they connect outward via combogenesis, and that will be particularly true during major events such as the transition to human culture.

As discussed, the larger group was necessary to provide numbers high enough for continuing complex culture and for innovations. These innovations came slowly at first, but then change sped up during various cultural eras. A given band did not have to rediscover tools and language because, like many animal societies, the band was potentially immortal. Yet because the bands were where people mostly lived and learned day to day, all three scales were necessary: the individuals (at the level of the complex multicellular organism), the band (at the level of the animal social group), and the new metagroup.

How about the new relations of the tribal metagroup? Let's consider it as a phenomenon.

First, unlike the band itself, the tribal metagroup has the ability to expand in size. This quality is related to its fuzziness. Cultural traditions have no inherent limit on extent. The cultural metagroup is at first simple, with lots of roughly duplicate, replaceable bands inside. And those bands can increase in number across the landscape. But as a general human social pattern that evolves, the metagroup also can and will get more complex internally—for example, in two levels hence with the geopolitical state.

Next, a metagroup inherently will face interactions with other metagroups. Those other groups are possibly different language groups, which to some extent keep people apart. Conflicts are of course in the works. But as culture expands, exchanges of rare raw materials and finished artifacts are crucial, and so is copying what has been witnessed or found. (I always wonder what Paleolithic people thought when they visited a cave with fine art on the walls that had been painted long before.)

Next, I believe we must consider metaphysical transfers in the new relations allowed by the metagroup. I'm not getting wishy-washy mystical here. I mean that humans, within a cultural matrix, now use imaginations to bring ideas into the culture. This is related to their capability of mental time travel. Thus, the new relations of this level's things are not just at the fuzzy, fluctuating, and potentially indefinitely expansive landscape borders of the metagroup. The "new" that is getting invented from the minds inside the metagroup is also part of the metagroup's relations. We should also include extinctions of cultural products here as parts of traditions get lost or are selected against by the process of cultural evolution.

New patterns of culture propagate in language, in artifacts and in rituals using artifacts, and in repeated behavioral patterns (chanting, then drinking, then dancing, for example): a metabolism of the "we." The "we" also allows people in and out. People can decide who is in and who is out with more flexibility and group-created nuance than can, say, wolf packs, as described in the previous chapter.

Giving names to things is an important way to bring material objects or people "in." What the Stanford archaeologist Ian Hodder has called "entanglements" is also relevant here.[32] The possibility, perhaps inevitability, that the entanglements will enrich themselves after this transition of combogenesis comes about from the nearly infinite kinds of transfers into the cultural systems. This enrichment involves the multiple scales of sharing now possible in the metagroup systems. Scholars have indeed focused on a new form of sharing[33] or exchange.[34] Thus, the new relations include, in growing number and complexity, everything that can pass across the fuzzy, undulating borders of the entangled, expandable, multiscaled "we."

The animal social group of the previous level hinted at these capabilities because individual animals could join or leave, and the animal group was formed by shared recognition of belonging. Now, with the human metagroup, materials and ideas can come in along with individuals. All those things can be discussed in the group. Such discussion forges explicit conformity (at least roughly) in the minds of people who share an understanding of what the group possesses.

How did this transition happen? The big-game hunting noted by Boehm as linked to a new kind of social egalitarianism meant that ancient humans were able to utilize a new type of food resource. Securing life in that challenging new ecological niche, which was not a simple kind of resource accessible to juveniles, required a highly integrated and ultimately larger social network. Gathering plant resources in massive amounts was crucial, too. The larger network of sharing and cooperation was enabled by shifts in mating patterns, toward monogamy and long periods in which siblings lived together (lifespan increased). A key innovation was accepting "in-laws" into one's cooperative network, even when siblings were split up by moving into other bands in the network. With these metagroup dynamics, cumulative culture could appear and flourish.[35]

The result: a big change from the relatively simple imports and exports of matter and energy of living things. And that innovation of biological import–export relations had been a dramatic shift from the static kinds of bonds that the things of the first five levels had, from quarks up to molecules. Now we apparently have a quite revolutionary new level based on its complex, open-ended relations that involve everything going on in the cultural metagroups. I suggest that vital to this discussion is humans' ancient ability to use the word *we* at a minimum of two scales in the tribal metagroup:[36] in reference to their band and in reference to the larger "people" or "tribe" (which includes other bands, which also can have names).

■　■　■

This chapter looks at the origin of culture from the perspective of combogenesis. Therefore, the emphasis is on building new relationships from things on previous levels. I believe I have made the case that the tribal metagroup is the new thing or system created on this level.

Note that with the operation of cultural evolution, a group itself can be subject to change by that evolution, to the extent that people can discuss individuals in relationship to the group as thing, and they can have opinions about the group's structure, at whatever scale. Those opinions can even be about the relationship between a smaller group and a larger one in the social nesting. As noted, language is key here and a primary enabler of such subtleties across the scales of the metagroup. Thus, the "we" itself—in at least two size scales—will be subject to further cultural evolution.

Finally, one might be tempted to characterize this level as things that now have ongoing, continuous, "micro-combogenesis." So much happens in the webs of exchanges. With cumulative culture, humans are always bringing things "in." As described earlier, there are numerous ways this "in" can be considered. Words that apply to anything from material objects to social rules are members that can come and go in these systems. Consciousness is integral to the picture of what is going on. People hold onto and live in mental maps, not just in physical territories. There are concepts of things, not just things. Thus, to distinguish subsequent levels of combogenesis—I suggest there are two more—then we might have to search for more than simply an ongoing normal process of addition of new materials, people, or ideas. We might have to search for truly outstanding transitions of the metagroup system.

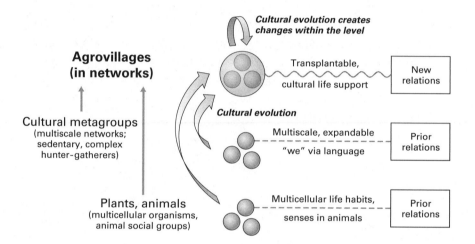

FIGURE 13.1

Networks of agrovillages formed when human metagroups combined with plants and animals from the prior levels of complex multicellular organisms and even animal social groups. The new life-support systems include plant and animal domestication. Cultural evolution is operating. (Included within the level of the cultural metagroup are the various types of hunter-gatherers.)

13

TRANSPLANTABLE AGROVILLAGES

SUMMARY: After the creation of human cultural, tribal meta-groups, do further events of combogenesis take place? Proposed here as a next level in the grand sequence is the domestication of plants and animals. These new systems of life support result from an integration of people with other species to create specialized "organs" that capture energy within those new systems. The novel types of networks enabled by this level can literally put down roots at sites across a vast landscape. Being transplantable, the webbing of agrovillages can in principle spread across much of the world at high density.

ANY NEW EVENTS OF COMBOGENESIS?

With the origin of the tribal metagroups at the previous level, humans gained the potential to live in networks of increasingly complex, multiple nestings of "we." The cumulative nature of cultural evolution and the lack of definite borders for metagroups made possible many avenues for growth. For example, the sizes of populations at nodes, formerly mobile bands, could increase. The scales and internal degrees of nestedness of the metagroups could grow. And the complexity of culture's entanglements could ramify with symbols and artifacts circulating within and among the nodes. What does this potential for greater complexity mean for distinguishing the next events of combogenesis?

This is a crucial question to ponder. We might entertain the idea that all further social and cultural development—from the caves and carvers of Paleolithic female figurines to the cities and architects of ancient Sumer and even to modern New York—took place within a single level. Vast changes occurred in those many tens of millennia, for sure. But perhaps the entities that resulted from those histories developed or evolved within a given level of the grand sequence. Changes, yes. Complexification, yes. More sophistication, depending on how defined, yes. But combogenesis? Please bear with me.

Let's illustrate this question with an example from biology: worms to whales. The earliest multicellular animals might have been soft-bodied, tiny tubular things living in marine muds between 500 million and a billion years ago. Eventually came the whale—for instance, the mighty blue whale, the largest mammal that has ever lived. It lives today. All this evolution in size, to be sure, took place within a single type of multicellularity: in this case *within* the biological clade of animals. However, many animals evolved to participate in various types of animal societies, and therefore the buildup *from* the animal *to* the animal social group is between two levels. (Reminder: social groups—say, of whales—then form contexts for further evolution of the animals—whales—within.)

Because in this book we have an ongoing interest in the distinction between changes *within a level* and changes *from one level to the next*, we must inquire into the nature of changes that followed the invention of the first human metagroups as described in the previous chapter. At that chapter's conclusion, I suggested that the metagroups are seemingly systems of ongoing "micro-combogenesis" in the sense that they are always open to bringing in resources, inventing new tools, adding words to language, adjusting rituals. Perhaps many thousands of years of changes in human social and technological networks should be considered all on the same level. If so, worm evolving to be whale would be analogous to the Paleolithic cave art site of Lascoux evolving to be Paris.

If a case is to be made for another *new* level along the grand sequence, one after the innovation of the cultural metagroup, then we need to find an event of combination and integration that is quite extraordinary. For animals and animal social groups, the two levels are easy to specify because of the bounded bodies of animals in the larger groups. It is going to be trickier with the inherent, ongoing micro-combogenesis and fuzzier physical characteristics of the human metagroups. Furthermore, for any candidate event of combogenesis, we must propose new relations that are decidedly innovative.

In this chapter and the next, I do make the case that at least two more major events of combogenesis occurred after the one that originated the earliest human metagroups. The first event is the subject of this chapter. It centers on food.

SIZE BREAKOUTS IN COMPLEX HUNTER-GATHERER SOCIETIES

We all need food. For millions of years, perhaps ever since Ardi, Lucy, or Turkana Boy walked upright, ancestors in our hominin lineage, when not lazing or grazing in trees, had to go out and gather, scavenge, or hunt for food. For much of this time and in most places, the sizes of the groups within which people had primary, daily contact with each other were small bands. Why? Local scarcity of fruit, tubers, and game limited the local populations. By the era of the Middle to Upper Paleolithic, cultural networks of early social systems within the language-possessing meta-groups fostered exchanges across geographically widespread bands. But locally and daily, as described earlier, those bands were still small in size, roughly on the order of modern chimpanzee communities.

However, in regions richly endowed with natural biological bounty, humans developed more intensively occupied sites within the hunter-gatherer lifestyle. Anthropologists have termed these sites "complex hunter-gatherer societies."[1] In them, people were at least in part seden-tary. Their architecture of simple huts (or houses) pinned down local nodes of populations within the wider networks of other groups that were either sedentary or perhaps more mobile. And the complex hunter-gatherers were experimenting with more intricate social stratification.

Such complex hunter-gatherer societies are known from archaeologi-cal and anthropological studies and were spread across multiple conti-nents. They often developed where seafood was rich in nutrients and abundant. Examples include the village-dwelling peoples whose liveli-hoods were based on the seasonal clockwork of giant salmon runs along the Pacific coast of what is now known as North America. On the other side of that ocean, also nourished by the sea, ancient Japanese Jomon cultures even developed one of the world's oldest pottery and lacquered-woodwork traditions.[2] On the Mediterranean seacoast, seminomadic people built dwellings at the site Ohalo-II in today's Israel. Ohalo-II was a settlement established at the astonishingly early date of 25,000 years ago, in the Upper Paleolithic.[3]

The development of complex hunter-gatherer societies also occurred at sites more inland as well. Wild grains and herds of herbivores specially flourished in the Levant region of the eastern Mediterranean, which includes a swath of southern Turkey. And there in Turkey lies the mind-opening archaeology of Göbekli Tepe.[4] Digs have unearthed stones still upright in their original circles, raised 11,000 years ago, like a premonition for Stonehenge 6,000 years later. Snakes and other animals in raised relief wrap Göbekli Tepe's stone pillars. Archaeologists envision periodic ritual gatherings at these circles. Perhaps these gatherings helped to unify and solidify a tribal metagroup network of dispersed villages of complex hunter-gatherers. All this was likely just prior to agriculture. It's perhaps no coincidence that the earliest domestication of a wheat (einkorn) took place nearby and relatively near in time, though the story is still developing.[5]

The bottom line: before ancient people possessed rich fields of crops, they found ways to support life in villages larger than Paleolithic bands and with some degree of social stratification. To me, this advance in complexity, both socially (implied) and architecturally (in the monuments), suggests changes that stayed within the level of primary metagroups. In these innovations, I see no obvious combogenesis. Again, an apt analogy might be to the evolution of multicellular animals, such as from worms to whales. In the human story, "we" could get bigger and more complex, but it did so within the same level of cultural evolution and as essentially the same type of system.

MULTIPLE ORIGINS OF AGRICULTURE AND THE COMING TOGETHER OF MULTIPLE SPECIES

Then came the origin of agriculture. I suggest that this event can be seen from a new angle using the lens of combogenesis.

In archaeologists' parlance, the classical, umbrella term for this event is the *Neolithic revolution*. That term boldly points out this event's transformational aspect in ancient prehistory, before writing. I believe the Neolithic revolution can be seen as a transition to a new level in the grand sequence from quarks to culture by combogenesis. How so?

Before getting into this event and its repercussions, I must note that although for simplicity I use the general term *agriculture* or *agrovillage*, I always mean to include animal husbandry. In fact, the first domesticated species, from wolf ancestors, was likely the woof-woof, the dog.[6] In most places around the world, the domestication of plants and the domestication of animals were closely linked. The event of taming plants and the event of taming animals often took place in the same broad geographical region and roughly simultaneously (over, say, a few thousand years or less for a particular region).

What came together in this proposed event of combogenesis? Quite simply, plants and animals came to be controlled components or subsystems within the larger human metagroups' cultural life-support systems. Humans learned to carefully tend and even worship these relatives from the earlier level of multicellularity. In such sites, there was gradually less and less going out and gathering, scavenging, or killing when hungry or raising the village en masse to labor for fish when the streams were flush with salmon. One could instead walk out, kill the fatted sheep, or dig into the common store of grain harvests that had been harvested from plants grown from seeds planted and tended the previous season.

How did this revolution happen? The details are buried and mostly lost in time. But archaeological digs have led to scenarios in which accidents at first led to increasingly more conscious incorporation of plants as the seasonal cycles of planting, tending, and harvesting occupied the tenders' minds. Early on, having accidentally dropped seeds along walking paths on which they brought wild harvests back to their villages, people noticed new food-bearing plants along those paths or around their villages the next year. Today's gardeners call these accidental plants "volunteers."

Once conscious planting of seeds began, seeds that hung on more tenaciously in the grain clusters (heads) of the crops tended to be the seeds that made it to a processing station after harvest. So there was a serendipitous selection for seeds that clung on the heads until purposively shaken or beaten. This selection occurred in broadly similar ways with crops derived from wild grasses such as wheat, barley, millet, maize, and rice, despite differences in cultures around the world.[7] Improvements also took place during the domestication of plants for edible tubers or fruits, such as potatoes and squashes in the Americas. All these events and more that were required for conscious agriculture were possible given human mental abilities to do mental time travel and the ability for groups of "we" to incorporate plants into their communal social systems. Some would say the crops also domesticated us to take care of them.

The animals that were domesticated were those that basically were friendlier. Dogs hung around camps, helped as signal guards against predators, and got fed in return. A symbiosis happened. The dog eventually became a member "inside." Historians of animal husbandry have noted that for the most part the domesticated species came from animals in herds or highly social groups. Such animals were thus used to living in groups and to leaders. Some of them were particularly tamable. (Again, maybe they were the smarties that tamed us.) Breeding eventually made the animals more gentle. Agreeable was "in," fierceness "out." Cattle, pigs, goats, and other species merged into the human systems.

Networks of agrovillages likely originated in regions in which those complex hunter-gatherer societies developed partly in response to abundant and nutritious wild grains, such as the Levant. The Stanford archaeologist Ian Hodder notes grinding stones in the archaeological record prior to agriculture.[8] Bruce Smith of the Smithsonian Institution has emphasized the origins of agriculture as an extended process, beginning with humans gradually constructing their special "niche" in nature, for example, by using fire to alter the landscape mosaic of vegetation more to their liking, well before crops lay tended in fields.[9]

So how should we value the eventual fusing of the plants and animals into the human metagroup? Might it be just a slight extension of the use of creatures that had been going on long before—say, using a wild plant vine as cordage to bind tree sticks together for a hut? Why should we consider the event a combogenesis?

One telltale answer involves genetic changes. Domestication brought about changes in the genomes of plants and animals. Worldwide, there were similar transformations in certain classes of wild plants as they were being domesticated into crops. For instance, as noted, over time seeds more firmly stuck to grain heads. Seeds got larger and more nutritious. They became more digestible. Though some of these changes were at first more or less fortuitous outcomes, there was eventually full conscious selection of what to plant. This conscious selection started pathways in time toward genetic changes as uncovered in the archaeological finds. Gentler animals also meant genetic shifts, which co-occurred with physical alterations in size, reduced teeth sizes, and more.

As humans grew dependent on these other species, the crops and husbanded beasts became dependent on humans for tending and breeding and therefore were evolved and biologically changed within the metagroups and cultural evolution.

Agriculture did not start with giant irrigated fields. As noted, it did not happen overnight. Early, iffy, tiny ancient harvests at first might have only displaced the least-efficient, most economically marginal aspects of the local gathering and hunting. But as crops and domesticated animals became more weighty to the human diet, a threshold was eventually reached in which the new system took over in sites and spread.[10] It did not spread everywhere: hunter-gatherers still exist today, as we saw in chapter 12, though there are vanishingly few. But the agrovillages were part of a spreading wave whose amazing progress is studied archaeologically and is vital to today's world.

Furthermore, the Neolithic revolution happened in many separate wide-flung regions. The origins of agriculture were multifold. Archaeologists recognize about ten regions of invention. Bruce Smith, for example, locates one in the Middle East (the Levant), two in China, one in Africa, one in New Guinea, three in South America, one in Central America, and one in eastern North America.[11] Many sites had local main grains that became staples for calorie intake—for example, maize in Mexico, wheat in the Levant, millet and rice in China. Potatoes were domesticated in Peru–Bolivia. Agriculture thereafter blossomed at other sites, which were influenced by the spread of people, information, and the plants and animals themselves (brought with the people or traded) from the sites of invention. In these later sites, some of the same main grains, for example, took hold but were evolved into different cultivars. And new crops derived from local plant species were added to the recipes.[12]

What spread were not just seeds, not just agricultural peoples and their metagroup communities, but the concept of agriculture as well.

To summarize: an event of combogenesis is here. Agriculture came about from a coming together and integration of plants and animals with people and their prior cultural systems. These helper species more and more became members of "us," the metagroup. New species became parts of the cultural webs. In this symbiotic merger with humans, wild plants turned into crops, and wild animal prey turned into domesticated herds or small groups (depending on the species). This event formed a new thing in the grand sequence through what was essentially permanent merger. It was a fateful transition that involved combination and integration. Agriculture was a score for combogenesis.

TRANSPLANTABILITY

If a new level of thing was created, we should be able to designate significant new relations at this level. Ever after the dawn of culture, humans were incessantly incorporating external things into their cultural systems—the continuous process of micro-combogenesis that I have noted. Change was often incremental and slow. But the potential always existed for something bigger to happen. The domestication of crops and animals was no ordinary "micro" event but rather an enormous one.

To get to the core of what agriculture could do, I suggest here an analogy to the modern International Space Station that orbits Earth. Fields of photovoltaic panels branch as great, black silicon "leaves" from the station's air-tight pods. These fields capture solar energy and turn it into electrical power for the station's high-tech, high-altitude, life-supporting metabolism that nurtures the brave voyagers inside. Some are thoughtful enough to sing songs to us back on Earth.

Fields of crops around ancient agrovillages were functionally like the solar collectors of the International Space Station. Those long-ago human habitats, as management centers, were in essence extended out to vast expanses of "engineered," living solar collectors: the planes, or plains, of the green crops. The result was a new kind of control on the input of energy to the human cultural system. The origins of this control heralded a new relationship gained by humans with respect to the capture of energy from the environment. Biological food energy did not have to be gathered from wild forging sites of "outside" plants. Instead, with agriculture the primary energy input was the sunlight itself. With agriculture, the sunlight was converted into food energy by the green fields of crops now "inside" the cultural system.

In some ways, with the combogenesis into this level humans became more independent from the environment. They had mastery of the energy capture, and that mastery required continual development of their agrovillage subsystems, such as improved storage subsystems (for example, secure grain bins) and special ways to coordinate labor in space and time.

For animal husbandry, the story is somewhat different. But it, too, transferred energy from the environment into the cultural systems. Formerly, people harvested wild creatures. Then they herded and managed domesticated animals across wild pastures. Later, with domesticated pastures as well, the energy loop was closed even tighter. People could grow the plants to feed the animals. The result was that as the domesticated animals fed on plants, whether wild or cultivated, the flow of valued animal proteins and lipids was embodied inside the cultural system. Energy storage walked around on hoofs within the system.

By themselves, these new forms of gathering and controlling energy would perhaps not have been all that consequential for human social life compared to the complex hunter-gatherer societies already in place at special sites. Recall that those societies were restricted to locales naturally rich with wild foods. The real key to the combogenesis reached by the agrovillages—the dramatically new relations of the new systems—was the *transportability* of the advanced life-support systems. To use a gardening term, the new agrovillage systems could be *transplanted*.

To where? To just about anywhere with enough water and suitable soil. Local edible wild foods were no longer as important, at least in theory. The movable feast of the agrovillage lifestyle created a new type of ecological niche that "[could] be exported to areas outside the original heartland."[13] This niche could be spread out to collect the sun's energy and tuck that sunlight into crops that were consciously planted, tended, and harvested.

Consider, again, the International Space Station. The capture of solar energy allows astronauts to persist in an otherwise hostile environment. Agriculture allowed an analogous inhabiting of otherwise hostile landscapes.

Yes, a few people probably lived in those tough wild sites before agriculture. But agriculture allowed the societies with the social and technical architecture of the complex hunter-gatherers to settle and thrive at high densities in places where that modality would previously have been impossible. And the settlements themselves, as nodes in the cultural webs, could reach higher and higher densities. It took time. Irrigation later made settlements possible in dry regions with large rivers,[14] and huge populations blossomed.

■ ■ ■

It's fun to flip the logic of this level. Rather than humans capturing plants and animals, we might consider that humans were captured and tamed to help special species of plants and animals, in the same way that flowers put forth pollen for bees to help the plants spread. Hodder's term *entanglement* applies here, pointing to how we are dependent on things and bound to care for those things, whether mechanical or biological.[15] Writers such as Michael Pollan remind us that we should look from the crops' point of view.[16] In reality, we are talking about *integrated* systems arising in this event. Especially when biological or cultural evolution is rolling, all the components that go into an event of combogenesis are altered by that integration.

To emphasize: the social systems of the new level could move and could bring their crops and animals with them as specialized energy organs within their metagroup systems. Populations expanded. Ideas were transplanted, too. Thus, with control and production of seeds and other edible goodies, supported by conscious knowledge and growing social networks, the new cultural systems had the ability to intensify locally and, crucially, to transplant themselves (like seeds) across the wide landscape. The new systems prepared the way for the next event of combogenesis to come.

14

GEOPOLITICAL STATES, MASTERS OF ACQUISITION AND MERGER

SUMMARY: The final level (for now) is the birth, development, and deployment of large geopolitical states. These cultural systems construct themselves by acquiring and incorporating smaller political entities. Beginning about five thousand years ago, the origins of geopolitical states across multiple primary sites are fostered by and further enable innovations in bureaucracy and sanction or social control. Not only are the resulting geopolitical states larger and built by combining smaller entities, but they also possess a new power. They know how to keep on incorporating.

MULTIPLE ARISINGS OF THE ANCIENT GEOPOLITICAL STATE

Archaeologists recognize the origin of the geopolitical state as the second of the two great "revolutions" in ancient cultural lifestyles that led to our world today. After the integration of humans with crops and animals initiated and developed the previous level, the next revolution went "urban."

Most archaeologists refer to this event as the "origin of the state." The term *state* can include early city-states centrally controlled by either a small group of people (oligarchy) or a hereditary ruler (kingship, monarchy). The state as a very broad and generic type of human organization also can include later developments into empires. And political scientists talk about the nations of the world today as "states." They debate what

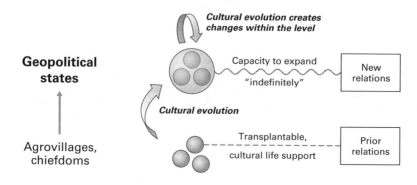

FIGURE 14.1

Geopolitical states formed in their sites of primary origin by the incorporation of chiefdoms or smaller states and the crucial innovation of expandable, hierarchical, social control (bureaucracy). "Indefinitely" is given in scare quotes because expansion is still limited by resources. Cultural evolution continues to operate. (Changes *within* the prior level of agrovillages created chiefdoms and multiple scales in the social networks.)

makes a state work or what leads to a failed state. To avoid confusion with other uses of this single-word term, such as for the fifty states of the United States, and because the ancient states were much about geographical extent (and territory is still so important today), I call the new thing or system of this proposed level the "geopolitical state."

Our first concern is with the beginnings of the geopolitical state at arrival on the new level. Here, the pioneer system is the first-generation or primary geopolitical state. Like the multiple origins of agriculture, this generation was born at multiple sites. The exact members of the set depend on the scholar. The list I give here is from Charles Spencer of the American Museum of Natural History in New York.[1] The dates are not moments a switch was flicked but rather the times that evidence shows a state to be in place and strong. The members, designated by region and capital city, are given in chronological order:

1. Mesopotamia, at Uruk (around 3500 B.C.E.)
2. Egypt, at Hierakonpolis (3400–3200 B.C.E.)
3. Indus Valley, in today's Pakistan, at Mohenjo-daro (2500 B.C.E.)

4. China, "probably" at Eritou (1700 B.C.E.)
5. Oaxaca Valley of Mexico, at Monte Albán (100 B.C.E.–200 C.E.)
6. Coastal Peru, at the Gallinazo Group (100 B.C.E.)

Large size was a common feature of these new metagroup systems, which developed from earlier, smaller units. For example, the city of Uruk likely held more than 50,000 people, Monte Albán more than 14,000. Around such urban centers sprawled a hierarchy of lower-tier, smaller-scale settlements. And the entire geopolitical state was not inside a political vacuum. Beyond its shifting and contested outer limits were chiefdoms large and small as well as even nascent states. Relationships with these outside polities might have been in the form of conflict or temporary treaty or trade. Possible successors to power, these outside polities potentially might often have been brought inside as new vassal polities at any time by the central state system's capabilities for expansion.

This ability to expand is key to this new level.

In their digs, archaeologists uncover certain keynote features of architecture. According to Kent Flannery and Joyce Marcus at the University of Michigan, the first-generation states had "palaces," "tombs with sumptuary goods appropriate for royalty," "standardized temples of a state religion," and "secular buildings whose ground plans reflect councils or assemblies."[2] Some standardized buildings and temples sprouted across a lower, second tier of smaller towns within the overall geographical extent of the state, but the state also had its central "Moscow" or "Washington, D.C." The result: you could walk down the streets of ancient Uruk five thousand years ago and feel more or less right at home or at least not wildly out of place, strolling among its throngs on the streets and in the markets, beholding spacious plazas with monumental architecture.

You must hope that in your walk your hands and feet were not bound to others marching in a line of newly captured slaves and that you behaved correctly in homage to robed and gold-bedecked priests. In cases used by Flannery and Marcus to derive common principles, such as the architecture mentioned, they also found that typically "state formation involved thousands of deaths, and thousands of other people were converted to slaves." They add, drily, "Sorry, but no one said that creating the first kingdoms would be pretty."[3] (They use the term *kingdom* for one of the early forms of the state.)

The primary sites of state formation developed out of regions that were more or less the independent origins of agriculture earlier. Thus, fitting the grand sequence, the level of agrovillages was a prerequisite for this proposed new level. And networks of agrovillages were essential components to the state. Gold-bedecked priests and fearsome rulers and their retinues need to eat. But is the state truly a new level according to the model of combogenesis? Sure, the larger scales of geography and settlement size are awesome, and complexity definitely intensified. But does the development of states indicate a new level or simply more of a continuum?

I have emphasized that events of combogenesis—especially for biology and culture—are not vertical snaps in time but rather ramps of developments. Yet for agriculture we faced a similar question. I hope I showed in the previous chapter that agrovillages did indeed mark a significant event of coming together and integration, especially with the new relations of transplantability. Now let's raise the issue again, using the same biological example from earlier. Are the geopolitical states, as networked systems larger than networked agrovillages, something (merely) akin to the way that whales might be compared to the simplest ancient worms—that is, an evolution in size and complexity but within the same level? Or are the geopolitical states new things that resulted from combogenesis and therefore on a new level and more like whales compared to their eukaryotic cells or complex social pods compared to individual whales?

STATE FORMATION AND TERRITORIAL EXPANSION

In southern Mexico spreads the broad, fertile valley of the greater region of Oaxaca. There, some two millennia ago, the formation of the Zapotec state took place with the central city Monte Albán. I have walked its plaza, climbed its stone pyramids, and reveled in the grand design that displayed through architecture this state's impressive political and social power.

Archaeological studies have shown that a decisive step was taken when the prestate chiefdom of Monte Albán fairly suddenly grew more powerful and conquered and absorbed three neighboring chiefdoms. A first-generation geopolitical state had emerged.

Charles Spencer, after studying the Zapotec and other first-generation geopolitical states in the list given earlier, concludes that "a common theme can be discerned: the emergence of each primary state was concurrent with the expansion of its political-economic control to areas that lay well beyond the home region."[4] Moreover, that expansion was to "areas that lay well *beyond a day's round trip* from the home region."[5]

The concept of a "day's round trip" will come into play shortly. But first I must note that Flannery and Marcus offer a similar conclusion. When they compare what happened in the Oaxaca valley of Mexico to the emergence of Uruk in Mesopotamia, perhaps the first state in the fertile crescent of the mighty Tigris and Euphrates Rivers, they say, "The Susiana plain of 5,500 years ago thus provides an analogy for what happened in the Oaxaca valley some 3,500 years later. Both regions had formerly been occupied by rival chiefly societies. The largest of the societies sought to take over the territories of the others, and eventually succeeded. The result in both cases was an early kingdom."[6]

Both Spencer and the Flannery–Marcus team emphasize expansion by takeover. The claim is that the first geopolitical states resulted from combination. That looks like a combogenesis. Is there an innovation in the result worthy of designating the resulting system as a new level?

Prior to this purported combogenesis into states were chiefdoms. According to Spencer and his colleague Elsa Redmond, the chiefdom is a "rank society ruled by a centralized and hereditary leadership, but its administration is not internally specialized."[7] So decision making, though centralized, was not yet bureaucratic. And the lack of bureaucracy or the existence of bureaucracy only in a primitive form led to a "built-in limiting factor."[8] Indeed, the territorial size of a politically bound web of villages in a chiefdom was around that time period limited to "a day's roundtrip" from the center. And a limit in size put a limit on complexity. In other words, the chief or the chief's representative could go out and enforce rules, check up on situations, listen to people. Certain innovations in the expanding delegation of power were not yet in place.

As we saw, hunter-gatherer societies could be simple or complex. In regions where food resources were especially abundant, the complex hunter-gatherers developed what anthropologists call "rank societies." There were chiefs, power struggles, even slaves. Across the world, there were clearly in place tendencies in human social evolution to find advantages in an ongoing exploration of getting larger and more complex (both at a social node—say, a village—and in the overall web of such nodes, the metagroup). We might even say that we see here a social-cultural mirroring of John Tyler Bonner's biological principle: "The top of the scale is always an open ecological niche."[9]

Thus, my call here is that size and complexity are not enough to form a new level (again, worm to whale evolved within the same level). With agriculture and its bringing of plants and animals into the cultural systems, the increase created a qualitative difference and a leap of advantages. It was an event of combogenesis. Once agrovillages arrived, the advantages continued for some locales to develop into larger and more complex nodes and networks. I take chiefdoms, therefore, on the same level as the agrovillages, like worm to whale. So is there anything fundamentally new about the state? I think the answer is yes.

Archaeologists have designated the state a revolution. Here I interpret this event through the perspective of combogenesis. *What is new is the new means of getting larger.* This new means is at the core of Spencer's mode of state formation by territorial expansion: the invention of *expandable bureaucracy* that enabled the new systems to go beyond prior limitations.

Benoît Dubreuil, theorist on the evolution of human hierarchies (some topic!), calls the development of the state probably "the most significant political change in human history."[10] In the path of size and complexity from egalitarian hunter-gatherers, there were always ongoing smaller innovations in what Dubreuil calls "social division of sanction." (Dubreuil tends to use the term *sanction* in its sense of punish [as a verb] or punishment [as a noun], related to the ability to enact or authorize penalties; but the concept can also apply to the ability to reward.) A chief in a chiefdom exerted more ability to sanction (punish) and thus exert control than did others. *But with the state came a major innovation*: the invention of the "delegation of sanction." Dubreuil says, "I used the concept of 'division of sanction' to refer to the relational mechanism through which specific individuals have the right to sanction a set of normative violations. The delegation of sanction goes further in that it entitles specific individuals to authorize others to sanction a set of normative violations. . . . I suggest that the emergence of the state follows from the expansion of this mechanism and the ensuing competition among organizations claiming a right to delegate sanction."[11]

This social innovation of the delegation of sanction was iterative and multiscaled. Appointees at a higher scale who could punish and reward could also appoint others to punish and reward at a lower scale, and they in turn could continue the iteration. The fractal structure could both contract or expand, as needed. The result was a vertically integrated and potentially growing hierarchy of attributes that Dubreuil calls "dependence and gratitude."[12] This invention of expandable bureaucracy enabled territorial expansion beyond previous limits. Now the political unit was capable of not just extraction—say, of material resources or people—from outside political units but also real takeover of them because the hierarchy of sanctioning could be expanded into the political units that were dominated and incorporated.

Here is Spencer's take: "The successful annexation of distant areas, those farther than a one-day round trip from the capital, required the leadership of an expanding polity to develop internal administrative specialization and the concomitant capacity to delegate partial authority to functionaries at distant outposts—in short, it had to bureaucratize."[13]

Flannery and Marcus also claim that internal social innovation went along with the ability to take over and incorporate on the large scale beyond the chiefdom. "The creators of first-generation kingdoms had no template to follow. They did not know that they were creating a new type of society; they simply thought that they were eliminating rivals and adding subordinates. Only later did they discover that they had created a realm so large that they would need new ways to administer it."[14]

The new systems were not just larger. There were strong positive feedbacks, in Spencer's terminology, between size and complexity, which developed in a concurrent, mutually reinforcing manner. The state's arrival, which Spencer says often took place relatively quickly (the formation), had new tricks. Some sort of leap happened. As noted earlier, this was not a flip-of-the-switch moment. But Spencer does claim the transition was sped along by the feedbacks. Dubreuil notes "the ensuing competition among organizations claiming a right to delegate sanction." Vying powers experimented with social structures. When one or more of them eventually hit on success, a positive feedback between size and bureaucracy led to the birth of the geopolitical states in the arenas of primary origin.

This analysis supports the idea of a major level created by combogenesis. The geopolitical states formed by combining. How they did that created a new mode of combining. This combogenesis—say, of chiefdoms or of lower, prestate polities—went along with discovery of a new method of political integration. It was a method, I suggest, that once in place created cultural systems on a new level of the grand sequence.

THE NEW RELATIONS: SEEING OTHERS AS POSSIBLE "US," WHETHER THEY WANT TO BE OR NOT

As always, for this phase of the grand-sequence logic, we want to ask about new relations gained and further developed by these new things, systems, entities, ontums. Does this level offer any interesting, significant new relations that prior levels did not possess?

New relations directly follow from the discussion in the previous section. They involve the issue of expansion and, specifically, expandability. At multiple first-generation sites around the world, the discovery that the innovation of bureaucracy via iterative, nested delegation of sanction could enable expansion and takeover led to a new view of the peoples and smaller polities who happened to be outside the state. Namely, those outsiders could potentially become insiders. And resources, whether food, goods, crafts, or people, could flow from the former outsiders toward the center with new ferociousness.

Spencer points out that once the state has been created, it "theoretically has the capacity to expand indefinitely."[15] Dubreuil concurs: with the innovations, the "indirect control of rulers over their subordinates' dependents paves the way to an indefinite growth in their political power."[16]

In agriculture, the new metagroups of agrovillages, in networks that included crops and animals, created high-density life-support systems that were transplantable. Now the expandable bureaucracy of the state—a new metagroup level—was also transplantable in that it could propagate while maintaining network ties to the center. The expandable bureaucracy took place both inside the main, close body of the state as its capital city grew and geographically outside it in the expanding network of territory.

Of course, the others outside often did not want to be insiders. Spencer notes "strategies of resistance."[17] For Dubreuil, "in practice, the expansion of the state will be limited by various factors, such as the presence of competing entrepreneurs or constraints on the extraction of the resources that are needed to support the relationships of dependents at each level of the hierarchy."[18] Thus came into play dynamics that could lead to takeovers of the state by the outsiders. Or from these dynamics the state might collapse in size into smaller units and then reemerge.

Archaeologists note "cycling" and subsequent generations of states, from secondary to tertiary, following their primary formation. Experiments! Some succeeded. Others failed. From failures and collapses came new attempts. Incorporation and networking took place by raw conquest and were augmented by social inventions that led to political and personal bonds—for instance, through royal marriages or through gifts such as dazzling gold jewelry that the central state could produce but outsiders could not. The overall result was the panorama of political developments of revolutions and evolutions of history, what Flannery and Marcus call "the chain reactions that create more kingdoms."[19]

■ ■ ■

Most often emphasized in studies of the geopolitical state are the practical drivers coming from what it gained, what the rulers achieved (at all scales of bureaucracy) in terms of material and human resources, and the biological fecundity of the people in the expanding populations. Mass actions could raise massive stoneworks, dig irrigation canals, support craft specialists. But we should consider the existential advantages of the geopolitical state, too. New scales of belonging came in, with shared symbols and meaning, one function of which was to manage the inner terror linked to the universal tolling of the bell of personal mortality.[20] Resulting features of the state included the formation of priesthoods as specialists in such management, who occupied impressive temple complexes and often claimed direct lines of access to supernatural ancestors or gods. Therefore, existential boons were crucial in the reckoning of what Flannery and Marcus call the "social logic" of the state.[21]

Much of this new social logic was, of course, conscious and could be consciously refined. The states could now consciously view outsiders as potential insiders and had the collectively conscious means to make that happen. Rulers could chest-thump their successes. The people could be consciously awed and grateful (but complicated emotions filled all!). Thus, the concepts needed to perform this state-style combogenesis could be repeatedly implemented and improved by the minds that were crafting the conscious aspects of cultural evolution.

Are we still in this level today? We might be. I revisit this issue at the end of the book. The territories of some ancient states were larger than those of many nations today. According to Flannery and Marcus, "a kingdom is one kind of state (almost certainly the earliest kind, with military dictatorships and parliamentary democracies . . . rising later)."[22] This idea is tantalizing. They imply that what seem to us different forms of government might instead be members within an inclusive, general array, which in the view of the grand sequence means an array that characterizes a level, here the new level of the innovation called the geopolitical state, first reached thousands of years ago. This was a significant new level in the grand sequence, and it is the final one I can at this point definitively identify.

3

DYNAMICAL REALMS
AND THEMES

———

The inquiry so far has produced a set of things to consider: twelve fundamental levels of a grand sequence, derived using combogenesis as an overall theme. Thus, the first aim of the book has been completed: to propose a unified approach to the sequential creation of our lived universe. The next aim is a quest for themes within the grand sequence.

Here a trio of dynamical realms in the grand sequence is proposed: physical laws, biological evolution, and cultural evolution. Each realm has a base level that initiated core dynamics for subsequent levels. These realms within the grand sequence are used to interpret generalized evolutionary dynamics, aided by a pattern I call the "alphakit": atomic, genetic, and linguistic. I also outline the big-picture themes or parallels between levels in the grand sequence. Finally, an epilogue asks if the findings can help frame what is happening in today's world, where a transformative future seems to be manifesting at an ever more rapid rate.

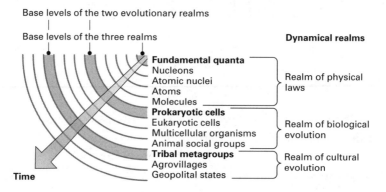

Base levels of the two evolutionary realms

Base levels of the three realms

Dynamical realms

Fundamental quanta
Nucleons
Atomic nuclei
Atoms
Molecules
Prokaryotic cells
Eukaryotic cells
Multicellular organisms
Animal social groups
Tribal metagroups
Agrovillages
Geopolital states

Realm of physical laws

Realm of biological evolution

Realm of cultural evolution

Time

FIGURE 15.1

The three dynamical realms, with particular spans across groups of levels. The base levels of the three dynamical realms are shaded in the concentric circles and are labeled in bold type in the list of levels. Note that two base levels (prokaryotic cells and tribal metagroups) initiate evolutionary realms.

15

DYNAMICAL REALMS AND THEIR BASE LEVELS

SUMMARY: The twelve levels form a set of similar phenomena that can be used to seek additional findings. Within the grand sequence, there are three largest-scale, natural families of adjacent levels, designated the dynamical realms: the realm of physical laws, the realm of biological evolution, and the realm of cultural evolution. Common themes shared by the levels within each realm are related to a realm's core dynamics. Within the grand sequence, the dynamical realms imply another theme: each realm had a first level, its base level, which can serve to focus further analysis.

SEEKING THEMES WITHIN THE GRAND SEQUENCE

Combogenesis has been the master theme of this book—the basis for the development of the grand sequence. The levels in the sequence share with each other the dual property that at least their initial members were generated by combogenesis *and* that some members of each level participated in a subsequent combogenesis. (Exceptions: the first and last levels are "single sided," but see the epilogue.) Thus, these levels constitute a set. They can be looked at as a "field."

Here we can explore common patterns—themes, parallels—within this field. We might probe the structure of the whole sequence by going back and forth among the levels, searching for themes that deepen our understanding of "everything" that lives in us and that we live within.

THREE GROUPS OF NEIGHBORING LEVELS

From the chapters in part 2, I repeat here the events of combogenesis and in italics identify by short phrases the most general dynamics of each event (see the figures that began these chapters).

- Fundamental quanta to nucleons (protons, neutrons). *Energy repose.*[1]
- Nucleons (protons, neutrons) to atomic nuclei. *Energy repose.*
- Atomic nuclei (plus electrons) to atoms. *Energy repose.*
- Atoms to molecules. *Energy repose.*
- Molecules to prokaryotic cells. *Transitional dynamics to biological evolution.*
- Prokaryotic cells to eukaryotic cells. *Biological evolution.*
- Eukaryotic cells to multicellular organisms—in particular, animals. *Biological evolution.*
- Animals to animal social groups. *Biological evolution.*
- Animal social groups to tribal metagroups. *Transitional dynamics to cultural evolution.*
- Tribal metagroups (plus plants and animals) to agrovillages. *Cultural evolution.*
- Agrovillages to geopolitical states. *Cultural evolution.*

I want to compare dynamics—basically, pattern formation during the interactions of things and relations—across levels from the perspective of combogenesis. Note that some italicized phrases that describe the event from one level to the next are exactly repeated. This repetition implies that dynamics of certain events can be grouped with others. Of course, these phrases, such as *energy repose* or *cultural evolution,* are repeated only because they are highly simplified here, in contrast to the details in the chapters that describe the formation of each unique level. But bear with me.

I propose that the twelve levels can be organized into three great groups or families of neighboring levels that share some profound similarity in their dynamics. These dynamics influence the mode of combogenesis at each event. But before specifying these three families, what should we do about the two events that are called "transitional dynamics" to a particular style of evolution?

For the proposed group of "biological evolution," I include the prokaryotic cell with its three subsequent levels that are governed by biological evolution. Yes, those first cells formed out of an earlier era of molecular, chemical evolution, and formal biological evolution was not "born" until those cells. But much of what the level of the prokaryotic cell is about, after its origin, comes from further evolution (a lot!) within that level. By a similar reasoning, I group the tribal metagroup level—which gave birth to full-on cultural evolution in our big-picture synthesis—with the two subsequent levels. We should not forget, though, that the prokaryotic cell and tribal metagroup are big transitions between basic classes of dynamics. That fact comes into play in chapters ahead.

We have, therefore, three main groups of levels. They are important in our search for themes. They form a framework for the search. As a suggestion, I call these three groups, families, or clusters the *dynamical realms.*

DYNAMICAL REALMS AND CORE DYNAMICS

A dynamical realm is a series of levels that share special, governing operations—that is, dynamics.[2] These dynamics are core processes of the workings of things and relations shared across the levels that constitute the realm. The implication is that spans of levels form categories larger than the individual levels themselves, promptly suggesting the existence of certain large-scale themes of the kind we seek. Thus, a dynamical realm is a kind of world or zone or space of behaviors. We see its workings appear when general aspects of explanatory logic repeat across levels for specific events of combogenesis.

The core processes under focus involve the methods of stabilizing things shared across various levels. For example, the biochemists Addy Pross and Robert Pascal discuss two broad "worlds" of things: inanimate and animate.[3] Those are at least roughly equivalent to the first two dynamical realms to be proposed here: physical laws and biological evolution. Jonas Salk, featured in the preface, designated three main types of "matter"—physical, living, and human. Those are very close to all three realms to be proposed here: physical laws, biological evolution, and cultural evolution. Furthermore, with core dynamics of realms as general ways of creating stability, we hearken back to the concept of stratified stability proposed by Jacob Bronowski, also featured in the preface.

We will explore how these core processes work in each realm as we get into the search and discovery of general themes in the grand sequence. Here I highlight a few features of these three realms, now formally named:

• *Dynamical realm of physical laws: the five levels from fundamental quanta to molecules.* The first types or things of each of these five levels formed in sequence as the universe cooled after the Big Bang. Further types formed in ordinary stars, the spectacular supernovas, and planets (for example, the molecule-based minerals). These processes are fairly well understood by physics and chemistry and do not require anything like biological evolution. We send probes to Mars and analyze its minerals, we write equations for fusion reactions inside stars, and we peer into deep space and witness the chemistry of interstellar gas clouds. The molecule is a special level in this realm, somewhat of a breakaway level, because the majority of its types are formed only in living cells, the next level. More about that later. Pross generalizes the processes of this realm as a "world" of "thermodynamic stability."[4]

• *Dynamical realm of biological evolution: the four levels from prokaryotic cells to animal social groups.* Now we encounter living things or systems that require imports of nutrients and energy and exports of wastes. Biological evolution applies. We saw some major aspects of shared logic in the explanations of combogenesis in the major transitions in this realm: the workings of evolution that had to deal with trade-offs and genetic conflicts of interest across scales. To happen, combination has to "pay off" in the support of life. The famous phrase of the biologist Theodosius Dobzhansky is relevant across the levels of this realm: "Nothing in biology makes sense except in the light of evolution."[5] Animal social groups are to some extent a breakaway level in that those groups in many cases have indistinct physical borders and their members, the animals, are not physically bonded. More about that later. All of these levels made their first appearances on Earth well before the start of culture.

• *Dynamical realm of cultural evolution: the three levels from tribal metagroups to geopolitical states.* With the start of these cultural levels, humans live on at least two scales of "we," aided by language, tools, symbols, and extended social networks. Terms such as *mind, consciousness, value, mental time travel, equality, obligations,* and *reputation* come into play in these systems. We see conscious decisions by people and among groups of people acting collectively. Decisions and collectives increase in scales of size and complexity with the levels. Cultural evolution is in place as an overarching operating principle. We might say, at least once language is turned on, "Nothing human makes sense except in the light of cultural evolution." Borders in these metagroup systems are not physically inherent but negotiable, and freedom from physical borders relates to the multiscale belonging in the primary metagroup structure that initiated this realm.

Now, it is true that each event of combogenesis engaged specific, new dynamics ("local" dynamics—though I do not use that term after this brief discussion). For example, the formation of animal social groups that were in our ancestry involved the dynamics of the senses in animals that use sight and sound. We could review examples of uniqueness at every level. Indeed, our look at the uniqueness of the new things at each level led to distinguishing the new "relations" of those new things. So in an attempt to keep concepts clear, I try to use the word *dynamics* exclusively in reference to the realms—in other words, for the overarching, shared operations across several levels. Thus, dynamics are aspects of things and relations fairly common across a realm's group of levels; in contrast, when I was laying out the levels, I emphasized the specific *relations* as new innovations possessed by each level's things.

NEW-REALM DYNAMICS INCORPORATE PRIOR-REALM DYNAMICS

These considerations point to certain crucial levels as ones that initiated new, overarching dynamics that continued as the major dynamics into some number of subsequent levels. I will discuss these special *base levels*, but first a clarification is needed.

This is important: The name of a realm extends from each of these crucial, initiating base levels to subsequent levels and up to but not including the next crucial initiating level. But with each next base level the dynamics of the previous realm are not negated but added to, incorporated, subsumed. Transitioning from one realm to the next does not mean that the dynamics of the previous realm get dropped or lost. But how the prior realm's dynamics operate in each next realm changes as they get incorporated in that new realm.[6]

For example, once physics and chemistry were operating, or, indeed, once the fundamental forces of the Standard Model came in, the forces of physics did not stop with the invention of the first living cells. The same goes for biological evolution. Biological evolution did not cease with the arrival of human culture. In fact, humans were able to domesticate plants and animals because those wild creatures had capabilities to evolve and to physically adapt to the needs of the agrovillage networks. Grain sizes got larger. Animals became more tame.

This feature of the realms—incorporation as they build—is like the nestedness of the levels themselves, but at a more general sweep of logic. A realm ends, in this proposal, when a level emerges in which the workings of its things are so innovative that it is worth designating not just a new level but also a new dynamical realm.[7] This happens at the transitional levels noted earlier. I call these transitional levels the realms' base levels.

Therefore:

- The laws of physics and chemistry extend from the fundamental quanta all the way to geopolitical states.
- The operations of biological evolution extend from prokaryotic cells all the way to geopolitical states.
- And, obviously, the workings of cultural evolution extend from tribal metagroups to geopolitical states (a span identical to the span of the realm itself).

Figure 15.2 shows the operation of dynamics for the three realms.

BASE LEVELS AND MAJOR DYNAMICAL TRANSITIONS

The existence of dynamical realms appears to be a large-scale theme within the grand sequence. As hinted at, another theme is immediately implied: each realm had a first level, a beginning. I gave a term *base level* for this pattern, but there were other possibilities, such as *initial*

Operation of dynamics:

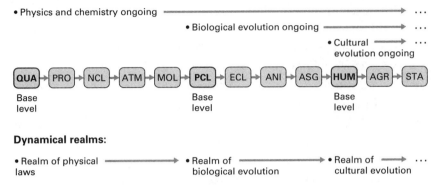

FIGURE 15.2

The dynamical realms are defined by clusters of levels, but the dynamics of each prior realm continue to operate even after a new realm begins. A base level starts each realm. The levels arc as follows: QUA (fundamental quanta), PRO (nucleons: protons, neutrons), NCL (atomic nuclei), ATM (atoms), MOL (molecules), PCL (prokaryotic cells), ECL (eukaryotic cells), ANI (multicellular organisms: animals), ASG (animal social groups), HUM (human tribal metagroups), AGR (agrovillages), STA (geopolitical states).

level, *birth level,* and *start level.* Base levels add a new set of fundamental dynamics to a previous set (keep in mind the discussion about incorporation).

Some events of combogenesis might be missing in the grand sequence as I have described it, especially, as discussed, during the origin of life. And I have said that new levels in biology and culture happen over a ramp in time, not in a vertical step that takes almost no time at all (from our human time perspective), as in reactions in physics. When scholarship confirms the features of any of these missing events, however, I personally think that we should be able to fit them into this book's overall conceptual framework of levels and realms (both appropriately modified) and that such modifications will be fascinating additions to our understanding of the progression from quarks to culture.

Now, a base level is more than simply the start of something new. What is new is powerfully potent. Each base level revs up a class of overarching, core dynamics that continues into subsequent levels of the realm and beyond—for example, the start of physics, the start of biological evolution, the start of cultural evolution. Thus, we can say there are innovations at each level, and at each new realm the innovations are exceptionally momentous—indeed, extraordinary. Therefore, again, any level that is purported to start a new realm implies that some very broad, persistent, new processes get initiated, not only for the things with new relations in that starter level but for the subsequent levels within the realm, up to the incorporation into the new realm. These exceptional base levels are (as also noted in bold type in figure 15.1):

- *The fundamental quanta.* Origin of the realm of physical laws.
- *Prokaryotic cells.* Origin of life and the realm of biological evolution.
- *Tribal metagroups.* Origin of language and the realm of cultural evolution.

In part 2, I covered what I consider to be the essentials of how these three special levels became levels, presenting each as just another level in the grand sequence. But now these three share a proposed theme: each is a base level to a dynamical realm. Because of what they share, we will revisit them all again, building from what was brought forward in their individual chapters and inquiring about the base level as an iterated pattern.

Furthermore, I suggest that the base levels serve as anchors around which additional comparisons might develop. How did each base level come to be such from the immediately prior levels of the prior realm? (Of course, this question does not apply to the fundamental quanta because there are no established levels prior to this one, but the question does apply to the starts of biological and cultural evolution.) And because biological and cultural evolution share what I call "evolutionary dynamics," they might contain some degree of similarity in the levels after their base levels. We shall see.

16

ALPHAKITS

━━━

Atomic, Genetic, Linguistic

SUMMARY: In the realm of physical laws, from quarks to molecules, there is a huge increase in diversity of types. A general pattern is developed: the alphakit. *An alphakit contains at minimum a pair of sets: a small set of basic elements and a expansive set of productions composed of those elements. Furthermore, both biology and culture possess levels that start new alphakits: the genetic alphakit in prokaryotic cells and the linguistic alphakit in human tribal metagroups, respectively. It is significant that these two levels are the base levels of the two evolutionary realms.*

DIVERSITY: FUNDAMENTAL QUANTA TO MOLECULES

Look at the base level of the realm of physical laws, the starting level of our universe. We ask how the fundamental quanta got to the point of yielding molecules, after a series of levels, and then to the point of a new dynamical realm. We know molecules are enormously diverse. But let's follow that process through the first five levels of the grand sequence.

The physicist Frank Close calls the fundamental quanta "the letters of Nature's alphabet, the basic pieces from which all can be constructed."[1] And that first level is also commonly called the "periodic table of particle physics." But, then, at the level of the atoms, the fourth level, we have what the New York University chemistry professors Trace Jordan and Neville Kallenbach call the "atomic alphabet."[2]

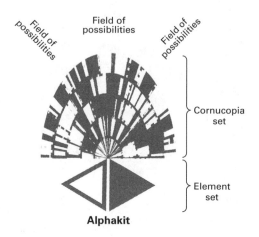

Alphakit

FIGURE 16.1

A generic alphakit, with its element set (represented here in black and white) and its cornucopia set (with a plentitude of types, represented by the permutations of bi-color patterns). The things in the cornucopia set can expand and contract in number of types (types of things are thus gained and lost, but there is usually a net gain). Expansion is possible even with a constant set of elements because the combinatorics of the things in the element set can create new possibilities, shown here as the field of possibilities.

So we have two alphabet metaphors here. Let's see how both can be right.

In the chapters for these levels (chapters 3–7), we saw that often the new relations were changes in magnitudes of the fundamental forces from the fundamental quanta and thus changes in how those forces interacted with each other, giving rise to the levels of what I now am calling the realm of physical laws. Here I want to review specifically the numbers of types at each level. I will then generalize to a pattern that I call the *alphakit*. This theme has additional relevance for biology and culture. To get that generalization, I ask a simple question for each level: How generative of types was that level for the subsequent level?

- *Base level of the entire grand sequence: fundamental quanta of the Standard Model.* Constituents (as types): Matter-field quanta of six quarks, six antiquarks, six leptons (including electrons), and their antileptons.

Force-field quanta of a handful of gauge bosons (gluons, photons, etc.). The numbers depend on how one counts (for example, some count the eight types of gluons). But that issue does not affect the answer to this question: How generative of diversity was this level for the next level? Not very: out of a hundred "species" in the particle zoo at the next level of quark–gluon combinations, only two (proton and neutrons) are parts of ordinary matter. As I said, the Standard Model was almost a dud. Fortunately, it wasn't.

• *Next level: nucleons (protons and neutrons).* How generative of diversity was this level for the next level? Fairly good. Protons and neutrons allowed the formation of ninety-two naturally occurring atomic nuclei (most from supernovas).

• *Next level: atomic nuclei (hydrogen to uranium).* How generative of diversity was this level for the next level? No increase. The atoms created by nuclei and electrons were within the spectrum of the same number of types generated at the previous level, ninety-two naturally occurring chemical elements. But that was OK! Organizing electrons into electric mandalas of surplus and lack around atomic nuclei was a giant innovation in relations, which gave rise to chemical bonding.

• *Next level: atoms (hydrogen to uranium).* How generative of diversity was this level of atoms for the next level of molecules? Gigantic, enormous! To be further discussed.

• *Next level: molecules (simple H_2 to complex, giant proteins).* How generative of diversity was this level for the next level? Next up from molecules is the origin of life, where the concept of countable types is not applicable in the same sense as it is for the levels of physical laws. The diversity of life is generated, and that's a stupendous feat. But the question of generativity has to be framed differently.

There are fine points in these counts. For both nuclei and atoms, for example, does one count all the isotopes? And does one include or not include the radioactive elements? I use the ninety-two naturally occurring chemical elements, which include a number of radioactive ones. But these questions are quibbles given our big-picture approach. They and other issues that might be raised do not alter a fundamental fact: going from the level of atoms to the level of molecules is an especially spectacular jump in terms of the resulting magnitude of the combinations.

Recall estimates for types of molecules: hundreds in gas clouds, thousands on planet Earth prior to life. But then, inside cells, we have Paul Falkowski's conservative estimate of "60 million to 100 million genes," and we assume at least one type of protein molecule per gene.[3] And then consider the more than 100 billion possible proteins in the "protein sequence space" or "protein universe." All from just ninety-two chemical elements in the atomic alphabet of atoms.

What about the quarks, electrons, and their quantized brethren as nature's letters? Well, the quanta did combine. And those combinations combined, and those new combinations in turn combined with electrons. At that point, the level of atoms and what they make, possibility took an enormous leap in diversity. So we might think of the starter set of fundamental quanta as something like "preletters" or "prealphabet," three levels down from the "letters" of the atoms. It took combogenesis three cycles of increasing nestedness to get from the starter quanta to a class of things whose relations were such that they really served as letters that could combine so prolifically that "words"—molecules—were possible in virtually uncountable numbers of types. Yet this combinatorial potential of atoms is nevertheless all thanks to the Standard Model.

INTRODUCING THE ALPHAKIT

The atom and molecule levels can be generalized into a pattern called the "alphakit." This theme deepens our insight into the next two realms as well, those of biological and cultural evolution.

The alphakit can be defined, at minimum, by a linked pair of sets of particular types of things (see figure 16.1).[4] Together, these two sets form a "kit" with inherent qualities of great potential utility:

1. *An element set.* An element set contains a small number of classes of discrete, relatively simple things. Consider the written Latin alphabet. Its letters, from *A* to *Z*, constitute one example of an element set. Other examples of element sets are the 1s and 0s of computer code, the twelve notes of Western music, and the numbers 0 to 9 in base-10 arithmetic. The element set might also be called an "array of elements."

2. *A cornucopia set.* In ancient Greek mythology, the cornucopia, or horn of plenty, was made from the horn of Almathea the goat, who suckled the infant god Zeus. The horn was later depicted overflowing with an abundant harvest. Thus, the cornucopia often symbolizes infinite abundance. Here I focus on the abundance of products in the cornucopia set. Perhaps think of the thousands of types of natural foods we enjoy, all from just a dozen or so chemical elements.

In an alphakit, the things in the cornucopia set are composites of the members of the element set. Thus, they are both larger in size and very much larger in number. Consider the tremendous number of words made from a finite number of letters of a written alphabet. Consider the awesome variety of music made from a small number of notes (though different octaves of the scale increase the number of notes, the notes are nevertheless tiny in number compared to the abundance of created music). Consider computer codes made from a simple binary system of 1s and 0s. Although the words, music, and computer codes, as cornucopia sets made from elements, are not actually infinite, they are richly plentiful for the complex workings and delightful creations of our cultures. For our purposes, they are innumerable.

To discern a pattern and say that it is an "alphakit," we thus require a minimum of these two sets of types of things: an element set and a cornucopia set. The exact numbers of types in these sets are not crucial, but their relative numbers definitely are. The huge increase in number of types from elements to cornucopia derives, of course, from the multiple ways to combine the elements. It's an amplification from the magic of combinatorics or permutations. (Not relevant here are the fine points regarding how mathematics formalizes those terms.) The potential yield from the permutating combinations can be quite large—gigantic, even astronomical.

One might be tempted to call just the set of elements the "kit." But because the two sets together form a system that is key to the workings of life in one important case and to the workings of culture in a second case, I prefer to consider the pair of sets together the "kit."

Also, let's note a third, invisible set. I mentioned the example of written words made from letters. What about the large number of potential words that do not yet exist? To avoid having too many terms named "sets," let me call this potential set the "field of possibilities." It lies outside an alphakit's attainments at any given time but makes a conscious nod to the fact that the cornucopia set can expand into possible attainments not currently in existence. Thus, the field of possibilities is typically very much larger than even the cornucopia set, which is very much larger than the basic and very small element set. The existence of a field of possibilities is important for evolutionary systems that might use alphakits to explore and create new things. Figure 16.1 summarizes these concepts.

Now, let's apply the concept of the alphakit to atoms and molecules. Clearly, atoms are an element set, the "atomic alphabet." Molecules are a cornucopia set with a field of possibilities even larger than their current tens of millions of classes. Let's call these two sets together the *atomic alphakit*. Keep in mind it's the atom–molecule pair of sets that is the "kit." With molecules, a cornucopia of creations arose. Molecules interact with others in chemical reactions and can both build and break into new types, exploring their *field of possibilities*. Life was eventually to be born from them. The atomic alphakit was key to this later event, as we have seen, but now this key has a name.

ALPHAKITS IN THE BASE LEVELS OF THE EVOLUTIONARY REALMS

With the theme of the alphakit, we turn to the base levels of the two evolutionary dynamical realms: the prokaryotic cell for biological evolution and the tribal metagroup for cultural evolution. Intriguingly, the new systems at both of these new levels contain new alphakits vital for the operations of those two things and for subsequent levels of the respective realms.

First, biology. The genetic code has won a deservedly starring role in understanding the operations of all cells, from the origin of life on. The code's four bases, A, C, T, and G, are often compared to letters of an alphabet, with genes as the words. Actually, the bases are weak direct generators of diversity because they make only sixty-four possible "codons" (taken as triplets, thus four possible bases in each slot, or $4 \times 4 \times 4 = 64$). Thus, the bases are more like, say, the nucleons (just two) that can make the ninety-two naturally occurring nuclei and then atoms. It's what the codons generate that is the true cornucopia set—the genes—typically consisting of a series of a hundred or more codons: thus, nature's 60 to 100 million genes.

In this system, which I call the *genetic alphakit*, there's a double alphakit. The codons directly code for twenty amino acids (some of the sixty-four codons redundantly code for the same amino acids, and I'm ignoring the fine point of "start" and "stop" codons). The twenty amino acids are strung together into proteins. Thus, life across our biosphere has 60 to 100 million proteins, at minimum. And the protein universe is even more "cornucopic," its field of possibilities essentially innumerable. The twenty amino acids are real stars here, as worthy of praise as the genes. They build into the cornucopia set of the workhorse molecules, the proteins. Given our two double-set stars here, it's best to considered the genetic alphakit as a double, connected pair of alphakits: (1) the codons and genes and (2) the amino acids and proteins.

Next, language. Language is complex, with multiple layers of nestedness and what linguists call the quality of recursion. But for our purposes we should start by considering the phonemes. Phonemes are elementary units of sound strung together to make words in oral speech.

English has about forty phonemes. Consider the fact that many vowels, though written as single letters—for instance, *a* and *i*—can be pronounced in various ways (as standard examples, the long versus short pronunciations of vowels we as kids loved to be tested on). This is the main reason why the number of phonemes is somewhat larger than the number of letters in the written English alphabet.

The number of phonemes in English can be debated, within a range. And the number varies with language. African languages, being more ancient, have more phonemes.[5] But the number of phonemes is small compared to the number of words the phonemes constitute. Although linguists can relate phonemes to classes of shapes of lips, tongue, and teeth as well as to gentle or puffy exhalations of air, the scale of phonemes themselves—as members of an element set—is nevertheless small and operates well enough to provide the central point: from a finite few an astronomical many can be generated.

Thus, phonemes are an element set; words are the cornucopia set. I call the "kit" of these two sets, so essential in our lives, the *linguistic alphakit*.[6]

A word about the cornucopia set of words in this kit. Back in time, after the origin of complex phoneme-based language, the naming of a thousand things became possible. How many words are we talking about? Shakespeare used 30,000. Most people actively employ several thousand in speech and can typically read 10,000. So the number of words is up a hundredfold or higher from the number of phonemes themselves. Given this kit's field of possibilities, James Joyce was able to coin the word *quarks*, which physicists then borrowed. Modern biologists, using numbers (0 to 9) to amplify the combinatorics of the alphabet, are naming the proteins as they find them. Thus, biologists can and with progress probably will have labels for all the proteins, even if millions of names are required. If a protein is there, it can be named. Talk about Adam naming the animals!

Many researchers have pointed out the analogy between genetics and language ("the genetic code is like language"). But formalized with the concept of the alphakit, this comparison might be more than an analogy. The analogy is a convenient way for a science journalist, say, to explain the genetic code by comparing it to language, but I think this similarity is worth considering as a much bigger pattern in the nature of existence than usually recognized. The analogy particularly shines forth as something to ponder in more detail when we place the alphakits into levels of the grand sequence. Frankly, it's astonishing.

Consider, now, the major innovations brought about by the genetic alphakit and the linguistic alphakit. In reference to the grand sequence, when did each start? According to the development described in this book, those starts are, respectively, at the origin of life with the prokaryotic cell and at the origin of culture with the tribal metagroups (the primary webs of multiscaled "we"). Those levels were also where the two styles of evolutionary dynamics began: biological and cultural.

A bundle of basic phenomena at the start of biological evolution was apparently rediscovered at the start of cultural evolution. This is what I find astonishing. Can the bundling of phenomena of base level, alphakit, and evolutionary dynamics be a coincidence? Can it be a coincidence that significant new types of alphakits came in with the things created in the events of combogenesis that also originated new evolutionary dynamics and thus started evolutionary realms? I think not. What do these parallels mean?

Before moving to consider this question further, let me deepen the parallel between the biological and cultural alphakits. Things in biology and culture, with their unique, innovative alphakits, have a series of higher-order nestings greater than even the cornucopia sets of proteins in the living cells and words in the cultural webs of "we." These higher-order networks derive from the interactions of the members of those cornucopia sets.

Biologists talk about proteins in systems between the scale of the proteins themselves and a living cell. Furthermore a cell includes other molecules such as water, special energy molecules, and lipids. But types of proteins in systems or networks do form the bulk of the working operations of life. Such networks include manufacturing loops such as the Calvin and Krebs cycles of energy transduction. Here we would include teams of proteins (proteins are never really alone) in both regulatory and signaling networks.

In culture, words are only rarely used alone. Right? Words usually make phrases and sentences; sentences build into higher-order packets of multiple, related sentences and from thence into narratives, stories, and sagas, such as the ancient, originally oral epic poem the *Iliad*. These higher-order combinations in language, which include what linguists call "recursion" ("Sally hit the ball that George had hit before"),[7] enable our complex mental constructions. Language in culture is a multi-scaled metabolism, like nested networks of protein systems in a cell. The networks of words are not culture's material complexity itself, of course, but the communication patterns of the networks correspond to the material and social patterns of the overall metabolism of culture. The nestings of language mimic (or model) the nestings of social relations and things in the "entanglements" between people and things, to use again the apt concept from archaeologist Ian Hodder.[8]

Thus, the members of the respective cornucopia sets (proteins [genes], words) of both genetic and linguistic alphakits are involved in higher-scale networks (protein [gene] networks, sentences, etc.) vital to the inner workings of the even larger respective things (prokaryotic cells, tribal metagroups) where those alphakits began. Now, with the parallel between the genetic alphakit and the linguistic alphakit more fully developed, we can take that parallel and consider the role of the alphakits in the base levels of the two evolutionary realms.

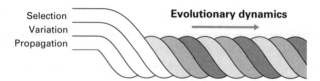

FIGURE 17.1

Evolutionary dynamics (in the general sense *not* limited to biology) consists of three component processes, depicted here as the interwoven strands of a braid: propagation, variation, and selection.

17

THEMES IN EVOLUTIONARY DYNAMICS

SUMMARY: Scholars have discerned core dynamics common to both biological and cultural evolution. Here these dynamics are called "evolutionary dynamics" and are conceptualized as three interwoven strands: propagation, variation, and selection. Starting at the base levels of the realms of biological and cultural evolution, alphakits and evolutionary dynamics are partners as themes crucial for the creation of subsequent levels. Furthermore, before those base levels came levels that had developed their own aspects of the themes: prior to the start of life are the atomic alphakit and chemical evolution; prior to the start of culture are social groups of animals with cognitive evolutionary dynamics. Enabled by language, full-on cultural evolution consists of a coupling between the more ancient cognitive evolutionary dynamics and the newer social evolutionary dynamics.

ALPHAKITS AND EVOLUTIONARY DYNAMICS
AT THE BASE LEVELS

We now examine an important partnership as a theme: between alphakits and evolutionary dynamics.

Consider the two kits described in the previous chapter: the genetic alphakit and the linguistic alphakit. When did each start? Again I emphasize that I am not trying to parse the details of the ramps in time for the births and bundlings of alphakits and evolutionary dynamics along different rates of development. I am only saying that at some point (a significant point!) they were strongly bundled and then remained coupled in subsequent levels. With that proviso, we have

- *Prokaryotic cells*: start of the genetic alphakit and the base level of biological evolution
- *Tribal metagroups*: start of the linguistic alphakit and the base level of cultural evolution

First, note that, once invented, the alphakits continued from the base level into the next levels. The alphakits in biology and culture were not just features possessed by the things of the base levels and then dropped in subsequent levels. It is true that features can disappear. The eukaryotic cells of your body, for example, are not bounded by the cell walls that bacteria or archaea have and presumably had throughout their ancestry. And unless you're on a survival reality show, I'm guessing you do not run your daily life by using the array of microliths our Upper Paleolithic ancestors used. Yet your cells do have the genetic alphakit, and you do use language. Thus, the alphakits of genetics and language, once started at the base levels, continued into subsequent levels.

This continuity makes sense. The kits are not easy to jettison without destroying the evolutionary dynamics they helped set into motion, as we will see. Both the element set and the cornucopia set in the kits were essential to the evolutionary dynamics that operated (and operate) to progressively change the things of those realms. Such changes included events of combogenesis to next levels in each of the evolutionary realms. And the continuation of each alphakit, once started, was a major factor in the overarching dynamics, which thereby helps us affirm that certain sets of adjacent levels are clusters or families worthy of the term *dynamical realms*.

But what about this term *evolution*, which I have been using for both biology and culture? In some sense, it's OK as a wishy-washy word for general "change." But that's not how I am deploying it. It is time to sharpen the term and show why it is relevant for our systems approach to biology and culture in the grand sequence, but not for any ol' brand of change.

First and briefly: biological evolution. Since Charles Darwin, the main principles of biological evolution have been well understood and increasingly refined—for instance, in comparative genetics and paleontology. Of course, Darwin did not know about the genetic alphakit. In chapter 8, I showed how the imports and exports of the prokaryotic cell virtually assured the trio of Darwinian processes, which are like the strands of a braid: propagation, variation, and selection.

CULTURAL EVOLUTIONARY DYNAMICS

Now for what deserves more unpacking: cultural evolution. A growing community of theorists and scholars have been pointing to certain common and fundamental principles in biology and culture.[1] Two of those scholars, Alberto Acerbi and Alex Mesoudi, point out that "cultural evolution studies are characterized by the notion that culture evolves to broadly Darwinian principles."[2] These Darwinian principles go by different names and are given different counts. I have seen as few as two and as many as six, but most often they are a trio.

So what are these Darwinian principles? The explications are quite consistent even though the short-hand terms vary among scholars. In this book, the trio is *propagation, variation,* and *selection*. Figure 17.1 illustrates the trio as an interwoven braid.

"Cultural evolutionists" certainly don't claim that everything in culture is explainable or under the control of an evolution-like process. But they do take the position that a generalized concept of evolution is as essential for inquiry into aspects of culture as it is for biology. I noted the trio of strands in chapter 12, on the topic of the tribal metagroup. I think the fact that these principles exist in culture is clear.

- *Propagation*: Analogous to the fact that life forms increase their numbers by reproduction, culture contains myriad types of propagation or increase (sometimes termed "copying" or "transmission"). People pass along stories and information; manufacturers create more and more "stuff" of all kinds; cities expand their grids; schools educate people to pass along skills and behaviors. In short, patterns of people, symbols, and artifacts are propagated, which almost always means at least the potential for the patterns to grow in number.

- *Variation*: One of Darwin's own "laws" was "variability."[3] Like propagation, variation is rampant in culture. People vary biologically, so we might expect their ideas, gossip, and preferences to vary. But we would also include here newly released models of cars, tweaks or even major shifts to a god concept, experiments in science, new products and market testing, sketches for a project in an artist's notepad, brainstorming in groups, doodling around in music, mulling over options for a vacation in one's mind. There are many scales of variation in all of culture's propagated patterns. Especially in our high-tech era, variation is often lionized as a goal to be sought.
- *Selection*: Biologists often separate natural selection, specifically as it relates to organisms' life-and-death struggles, from sexual selection, which involves mate choice. Selection is multifaceted in culture, too— even more so. We would include here mass selection by market forces on things, ideas, and even people (democracy). Selection is going on in minds all the time, via choices among cognitive scenarios. Indeed, decision making on all scales is crucial, including for the shaping of law, hiring and firing, college acceptances, what to stock in stores, advertising development, design from microchips to skyscrapers, editing of newspapers and blogs—from minor decisions about lunch to major ones about whether to launch billion-dollar projects. Variation and selection are often formerly coupled in how all manner of innovation is created.

William Calvin, scholar of generalized evolution, notes that both biological and cultural evolution share the fact of limited "workspace."[4] This "workspace" can be material or mental. A meadow has limited space for species, with constraints from sunlight and water. Humans sitting around a campfire have a limited capacity to hold stories in their heads. Thus, in both cases selection occurs. It's important that selection that contributes to the evolutionary shaping of pattern is not random. The biologist David Sloan Wilson emphasizes the "consequences" of the variants.[5]

The operation of these three principles, whatever the substrate, creates adaptive patterns by an evolutionary process. In Darwin's immortal words, as the "laws" play out in the braided trio of principles, "endless forms most beautiful and most wonderful have been, and are being, evolved." Those words can apply to both biology and culture. An eagle is like a successful equation, a symphony like a neuron.

What to call this expanded concept of evolution? Just the word *evolution* can be confusing, unless always qualified, because of its history of referring specifically to biological evolution. Another problem, as noted, is that the term *evolution* is too often applied very broadly to almost any change. Astronomers often loosely talk about the "evolution of galaxies," and even my closest colleagues in earth science talk about the "evolution of Earth," even though there was no selection of the planet based on any consequences from how well it performed in a population of variant planets.

The scholars who generalize evolution into the cultural realm have developed various terms for the interwoven trio I have called "propagation," "variation," and "selection" here: *logical skeleton, Darwin machine, algorithm, engine, formula, heuristic,* and even *recipe* (which I like, with its "ingredients").[6] The overall paradigm has been called "universal selectionism," "universal Darwinism," or simply "selectionism." Are we witnessing cultural evolution in action here in the search for a term?

For this book, my choice is the term *evolutionary dynamics.* I hope with the adjective *evolutionary* modifying the noun *dynamics,* we sail away from some of the shoals of misunderstanding surrounding the solo word *evolution* and emphasize evolutionary dynamics as a class of systems that includes more members than just biological evolution. Indeed, in a bit I need to bring in "chemical evolution" and "cognitive evolutionary dynamics" as other members of this general class.

PARTNERSHIP: EVOLUTIONARY DYNAMICS AND ALPHAKITS

Now what does the alphakit have to do with generalized evolutionary dynamics? Earlier I noted a bundling and referred to a partnership. What is this partnership?

Quite simply, it appears that the alphakit comes into play as a functioning part of things in biological and cultural evolution. To see how, we can use the three principles (strands of the braid) to see how the alphakit aids them. Biological evolution has a partnership with its genetic alphakit, cultural evolution with its linguistic alphakit. Can we say anything general about these partnerships? How might each strand—propagation, variation, and selection—be helped by the alphakit?

- *Propagation and alphakits*: In complex, propagating systems, such as the prokaryotic cell and tribal metagroup, an alphakit makes it easy to generate complexity (the cornucopia set) by simple means (the element set). Thus, the alphakit promotes construction of complex systems, and because the elements change relatively slowly, the alphakit helps ensure fidelity of the complex patterns being propagated. For example, the small set of twenty amino acids can be kept constant and has been constant for all life forms (with few exceptions) presumably for billions of years since the origin of life and LUCA, our last universal common ancestor. The small set of phonemes in language is relatively easy to use as a common base that carries on across generations in the cultural metabolism that relies on complex language. Yes, the phonemes themselves can and do change (they vary among languages), but they remain elemental particles of speech and much more stable in time than words. You can come up with a new word, such as *combogenesis* or *alphakit*, and hope that it spreads (or at least have it propagate as useful in your own mind). But good luck trying that with a new phoneme.

- *Variation and alphakits*: The alphakit can also contribute in a straightforward way to variability, because simple changes in the arrangements of the elements can create new members of the cornucopia. In biology, the variability is seen in mutations. There are many kinds of mutations, but the most basic type involves a switch of amino acid (DNA bases switch, making new codons and thus in many cases coding for different amino acids). In language, different words vary phonemes, as in *ball, tall, bill, till*. The field of possibilities around an alphakit means new combinations are also "out there." In biology, that implies a vast field of possible protein sequences from permutations of amino acids. In culture, many more words are possible than are at any time in use by tongues. When there is a need, it is easy for anyone to make up new words, such as *quark, gene, Beatles*, or *meme*. The types of variations in biology and culture are both complicated and multiscaled. The alphakit thus provides a structured way to explore the field of possibilities in a manner that is both controlled and prolific.

- *Selection and alphakits:* In biology and culture, given the fact that not all variants can indefinitely or infinitely propagate, selection follows. Relevant here is Calvin's concept of a limited "workspace," whether material or mental. Although there is nothing inherent in the alphakit that contributes to this third strand of the braid, the conclusion that an alphakit does directly aid both propagation and variation means that it also helps create selection as a unavoidable consequence. I have more to say about this connection later, specifically pertinent to language.

In summary, the alphakit is a way to aid faithful propagation of very complex systems of living cells and human tribal metagroups. The alphakit is a way to both explore and control variations. And, as a consequence, the alphakit is instrumental to selection because selection follows from propagation and variation. Figure 17.2 shows the binary partnership between the alphakit and the braided trio of principles.

From this discussion, it seems reasonable that for each evolutionary realm, a closely meshed partnership formed between a new alphakit and a new mode of evolutionary dynamics. Such a partnership entered the grand sequence at the special base levels of the realms of biological and cultural evolution. The coupling of alphakit and evolutionary dynamics in those realms continued as crucial in the levels after the special base levels.

Also, note that in the grand sequence a second major mode of evolutionary dynamics (culture) came from the earlier mode of similar dynamics (biology). Because biological evolution generates patterns, generating a new type of evolutionary dynamics became a possibility. And generating a new form of alphakit also became a possibility from biological evolution. We will soon look into this connection further. Of interest is the

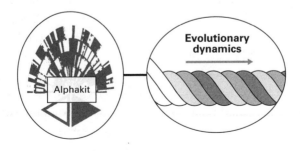

FIGURE 17.2

The alphakit and evolutionary dynamics operate as a partnership.

fact that the partnership—in its abstracted, pattern sense—was born twice, as if born once and then echoed or paralleled.

I am making the case here that alphakits and evolutionary dynamics are important themes in the grand sequence. Base levels and realms are also themes. There is more to find, I submit, from additional focus on all these themes.

ALPHAKITS AND EVOLUTIONARY DYNAMICS PRIOR TO THE ORIGIN OF LIFE

Do the levels immediately preceding the two evolutionary base levels have common features? Those preceding levels were like launch pads that gave rise to the evolutionary rockets of the base levels that followed. Here we use the themes of alphakit and evolutionary dynamics to explore and compare these transitions. In this section, I discuss the first case: molecules leading to prokaryotic cells.

Molecules are the cornucopia set of the atomic alphakit described in chapter 16. The complex interactions of molecules were able to create cells at the origin of life and biological evolution. Presumably, and with a respectful nod to the unknowns, systems of chemically reacting and mutually creating molecules underwent "replicative chemistry."

I also note, relying on experiments and theory alluded to in chapter 8, an ancient era of "chemical evolution" within the level of molecules. This chemical evolution would have ratcheted toward more complex and stable structures and toward the emergence of life.[7] As I have said, these self-replicating molecular systems could and probably will eventually be considered a level in between molecules and prokaryotic cells. This chemical evolution would be another type in the general class of evolutionary dynamics, different in specifics from full biological evolution and also of course from cultural evolution. Such a shift in modes of evolution from chemical to biological would have been even more profound than changes to biological evolution as it emerged, but those biological changes can serve as a model for envisioning the fact that modes of evolution can themselves evolve.[8]

Figure 17.3 puts these concepts together. The point is that this development from the earliest chemical evolution through phases of biological

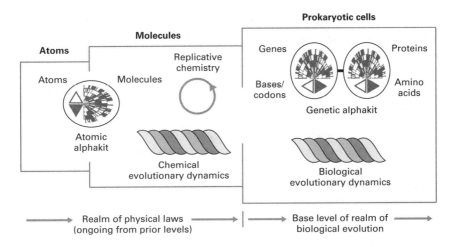

FIGURE 17.3

Both evolutionary dynamics and alphakits play a role in the transition from physical laws to biological evolution. The atomic alphakit manifests in the two levels of atoms and molecules. The cornucopia set of molecules and the ancient era of chemical evolution give rise to life, with new, biological evolutionary dynamics. The genetic alphakit consists of two element sets (bases/codons and amino acids) and two cornucopia sets (genes and proteins).

evolution began with the vast capabilities of molecules to manifest an astronomical number of types and for atoms to serve as a small set of elemental building blocks for all those types. Now, what about our themes?

The existence of the crucially important, molecule-based genetic alphakit (a double "kit," as shown in figure 17.3) inside prokaryotic cells was enabled by the fact that the preceding two levels in the realm of physical laws were actual sets of a different alphakit: the atomic alphakit of atoms and molecules. Indeed, the atomic alphakit was an active participant in forming the relationships that led to the later, genetic alphakit and even to the much later linguistic alphakit, given that humans are molecule-requiring beings who think by using neurons coursing with molecular and atomic interactions.

Furthermore, the emergence of biological evolution from the realm of physical laws came about because chemical evolutionary dynamics were made possible by the atomic alphakit at the final two levels of that physical realm. Thus, one might think of those final two levels—atoms and molecules—as a metaphoric "graduation" to the start of a new realm. That graduation was able to take place because the key patterns or themes—alphakit and evolutionary dynamics—so prominent at the start of the new realm of biological evolution had come into existence in different manifestations at the end of the realm of physical laws. Again, these repeated patterns are shown in figure 17.3. Atoms and molecules did not of course "know" they were ending a dynamical realm. It's just that the patterns they had arrived at were, in a sense, a pad from which the new realm of biological evolution could be launched with a reinvention of the same patterns in new forms.

COGNITIVE EVOLUTIONARY DYNAMICS PRIOR TO THE ORIGIN OF CULTURE

We now turn to the animals and animal social groups as the levels that preceded and led to the transition to the human tribal metagroups and the start of cultural evolution. I will show that another type in the general class of evolutionary dynamics began within those preceding levels. These "cognitive evolutionary dynamics" were important in the emergence of cultural evolutionary dynamics.

It's a given that the forces of biological evolution create functional adaptations. Some of those adaptations *within* living things can be "evolution-like." One recognized example is the adaptive, evolution-like immune system of vertebrates.[9] Another is learning in animals.[10]

According to Clive Wynne, specialist in animal psychology, for the type of learning called "instrumental learning" "the analogy to evolution appears to be obvious. An individual confronted with a problem generates a set of variant responses; these are tested against the problem and the most successful form the basis for a new set of variant responses. As conditioning proceeds[,] the variability of behavior usually declines until a single most efficient solution behavior (or a small set) remains. . . . There is variation followed by selection leading to a gradually changing population of variants."[11]

I call this inner evolutionary process within the brains or minds of some animals (surely our primate ancestors) *cognitive evolutionary dy-*

namics. All we need to accept here is that there are neural and behavioral operations that contain correspondences to the trio of logical strands of propagation, variation, and selection. There is no claim here about conscious versus unconscious[12] or any delving into the devilish details of kinds of learning. Wynne, for example, points out that whereas in biological evolution the strand of selection takes place among a pool of coexisting variants, in the learning that he describes the variants occur sequentially in the behavior or mind, as you like. Again, here we need only the fact that there is some form of evolution-like dynamics going on in cognition.

CULTURE: DUAL SCALE—COGNITIVE AND SOCIAL EVOLUTIONARY DYNAMICS

How did this prehuman animal foundation get to humans and cumulative culture as a new realm of evolutionary dynamics? Many researchers are working on that question.[13] Fortunately, a detailed answer is not required for us to proceed with a path of logic that involves major patterns within the grand sequence. Figure 17.4 puts the relevant concepts together.

In the pair of levels—animals and animal social groups—a capacity, as an adaptation, for cognitive evolutionary dynamics was shaped in the animals by the forms that their social groups took. This shaping went in both directions, shown in figure 17.4 as feedbacks between group and individual cognition, discussed as social learning in chapter 11. Then, the prehuman animal's cognitive evolutionary dynamics clearly shot up in sophistication during human evolution, for example in human capabilities such as mental time travel. To form the level of the tribal metagroups, as I have defined it, one requirement was human language: an innovative kind of alphakit that rode on the sound waves that had been previously used among certain animals as relational bonds within their social groups. How does language tie into the more ancestral cognitive evolutionary dynamics of many animals?

Chimps use and experience affirmation and denial as part of social learning. In their fission–fusion social groups, chimps can express this dichotomy by grooming or by blows. The lives of primates, like those of

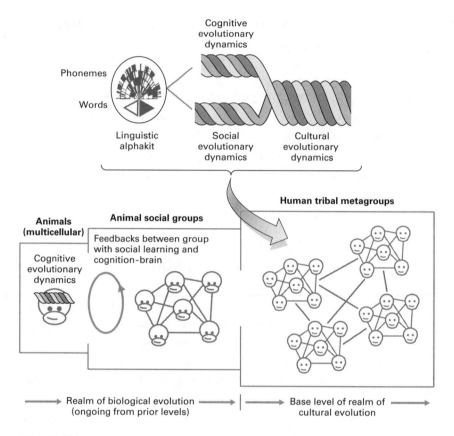

FIGURE 17.4

Evolutionary dynamics and alphakits in the transition from biological evolution to cultural evolution. A cognitive evolutionary dynamics involved with social learning of certain animals gets evolved into more sophisticated cognitive dynamics in people, who in the context of their tribal metagroups develop social evolutionary dynamics and the linguistic alphakit. Thus, cultural evolutionary dynamics has two components, cognitive and social—both are types of evolutionary dynamics and coupled.

most social mammals, are tangles of approval and disapproval. But with the arrival of human language the forms of acceptance and rejection ramified into new worlds of sophistication.

As shown earlier, language as an alphakit aids the strands of propagation and variation within cultural evolution. I said that selection follows (from the fact of limited mental or social "workspace"). But here I want to note further how language specifically enables intricate forms of selection in human culture.

The capability for language to be used for complex norms and strategies of "yes" and "no" is so important that the renowned German sociologist Niklas Luhmann termed the duo affirmation and denial a primary "binary coding" that language gives to social evolution. "Language offers a positive and a negative version of everything that is said." This coding "makes it possible to doubt what has been uttered, to refrain from accepting it, to explicitly reject it, and to express this reaction understandably, thus reintroducing it into the communication process. . . . Without binary coding, not even deferral would be possible, for we would be unable to recognize what is deferred."[14]

Note here how interwoven the strands become. Deferral for later communication is also a form of propagation, presumably with variation. Luhmann shows how this coding proceeds as a deep structure to guide the creation of moral codes, rituals, and religions. Eörs Szathmáry, a leading researcher in the major transitions of evolution, notes the result is a new type of open-ended creation: "It was language, with its unlimited hereditary potential, that opened up the possibility of open-ended cumulative cultural evolution . . . specific to humans."[15]

From exploring the presence of the themes of alphakits and evolutionary dynamics, we are led to a proposal. For certain animals with cognitive evolutionary dynamics their ongoing learning within sufficiently complex social systems required a "metabolism" of approval and disapproval among individuals. As biological evolution gave birth to cultural evolution, the linguistic alphakit was invented and was partnered with a new mode of evolutionary dynamics. Language allowed human metagroups to have fully developed cultural dynamics, which began a new dynamical realm. Complex, social decision making was a new feature. For example, the anthropologist Christopher Boehm has described members of a hunter-gatherer group sitting around and talking about whether a certain individual (not present) should be ostracized or not and, if not, what was to be done about the defector to restore approved behavior.[16]

This leads to the conclusion that what I had previously simply called "cultural evolutionary dynamics" actually consists of *a pair* of interacting, coupled subsystems that are types to be recognized within the overall general class of evolutionary dynamics. Shown in figure 17.4, the pair consists of the newer social evolutionary dynamics and an older cognitive evolutionary dynamics. The older one, again, existed in the brains of certain animals but then was tremendously evolved during human evolution with the coming of language and the added development of participation in social evolutionary dynamics with the protocols of collective decision making that was at least partly conscious (ha!) and language based.

Thus, the start of full-on cultural evolution consisted of two, coupled scales of evolutionary dynamics: cognitive within individuals and social among members of the tribal metagroups.[17] I also note that in addition to modulated sound, modulated light was important early on, too, for many researchers consider gestures to have been crucial in the early evolution of human language.[18]

ALPHAKITS AND EVOLUTIONARY DYNAMICS AS RECURRENT THEMES

Let me summarize what we have now seen about the two remarkable transitions in which one dynamical realm gave rise to a new dynamical realm. The focus is on the occurrence and roles of the alphakit and evolutionary dynamics as themes in relation to levels, base levels, and evolutionary realms.

• *Molecules expand up into prokaryotic cells (a base level).* This transition is from the realm of physical laws to the realm of biological evolution and thus *takes a nonevolutionary realm to an evolutionary realm.* With prokaryotic cells is born a partnership of the genetic alphakit and biological evolutionary dynamics. The preceding level of molecules supplied the cornucopia set in the atomic alphakit (consisting of the two levels of atoms and molecules). This cornucopia set was instrumental in generating (and containing) the complexity necessary for the start of life with the partnership just noted, and the set generated an era of chemical evolutionary dynamics within the interactions of molecules and prior to the origin of life.

• *Animal social groups expand up into human tribal metagroups (a base level).* This transition is from the realm of biological evolution to the realm of cultural evolution and thus *takes one evolutionary realm to a second, new evolutionary realm.* With tribal metagroups is born a partnership between a linguistic alphakit and two coupled subsystems of evolutionary dynamics (cognitive and social) that make the full-on system of cultural evolutionary dynamics. The preceding level of animal social groups supplied animals that had primitive yet sophisticated cognitive evolutionary dynamics, attuned to social learning in those groups. Those cognitive evolutionary dynamics were instrumental as a foundation for making the interactions among animals more complex, eventually giving rise to the linguistic alphakit as humans evolved, which added language-requiring, complex social evolutionary dynamics to the more ancestral cognitive evolutionary dynamics and the start of human culture as defined by the level of tribal metagroups.

Thus, we see that in the realms of physical laws and biological evolution their "graduation" levels or launch pads came about when an alphakit and/or evolutionary dynamics formed prior to the new alphakit–evolutionary dynamics partnership within the new things of the actual base levels of the evolutionary realms. In both cases, the graduation levels that culminated a realm are best understood as a pair. For the realm of physical laws, that pair was the atoms and molecules of the atomic alphakit, and the molecules underwent chemical evolutionary dynamics. The graduation, launch-pad level had an alphakit (especially the cornucopia set of molecules) and an evolutionary dynamics (chemical). For the realm of biological evolution, that pair was the animals and animal social groups, with the animal's biologically evolved cognitive evolutionary dynamics used for living in the context of the group.

There is some logic to the time–history within the grand sequence of these repeated themes or patterns:

The four levels of the realm of physical laws (quarks to atoms) that preceded the levels of molecules were not able to launch the evolutionary realm of biology. That happened only with the cornucopia of molecules, a set of things complex enough to give rise to an era of chemical evolutionary dynamics and then the genetic alphakit and the realm of biological evolution.

Biological evolution gave rise to cultural evolution by evolving as an adaptation a type of evolutionary dynamics first (animal learning of certain complexity). This makes sense because there is no problem conceiving of that learning as a functional adaptation that could be evolved. It just took the multicellular animal to reach that invention and the context of complex enough animal societies to eventually be able to expand into the human metagroups of cultural evolution. I leave it to further work to determine if an alphakit is in the brains of those animals—for example, in the form of a neural code. Certainly the on–off capability of neurons is like a binary code of some kind.

This discussion, I believe, reinforces the finding that certain patterns, such as alphakits and evolutionary dynamics, require special notice in understanding the big-picture interplay of fundamental classes of things and relations both within and between levels of the grand sequence.

18

CONVERGENT THEMES
OF COMBOGENESIS

SUMMARY: Here we inquire into parallels between the levels of eukaryotic cells and agrovillages, both "second" levels in their evolutionary realms. Their new things resulted from a merger of unlikes with complementary functions. In both cases, when the merger was complete, one side of the partnership became functionally specialized organs for enhanced energy operations. The main themes throughout the grand sequence are the Standard Model with its fundamental things and relations; alphakits in all three realms, including in the base levels of the evolutionary realms; evolutionary dynamics; and the existence of things and relations across the multiple scales of the grand sequence, with levels enabled by opportunities and often expressing parallel patterns during the rhythmic, iterative process of combogenesis. (See figure 18.1)

PARALLELS BETWEEN EUKARYOTIC
CELLS AND AGROVILLAGES

After the base levels of the two evolutionary realms came the "second" levels, the eukaryotic cell (biological evolution) and the agrovillages (cultural evolution). Given the parallels of crucial properties originating from the base levels, we might here compare these second levels as sites for additional parallels. Figure 18.2 shows their locations, emphasizing their positions after the base levels.

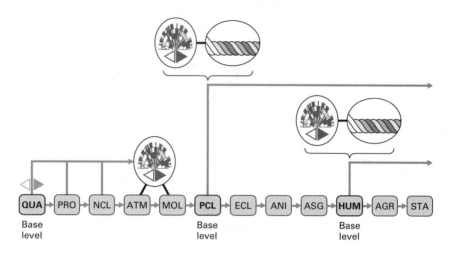

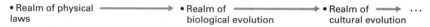

Dynamical realms:

- Realm of physical ──────────────────▶ • Realm of ──────────────────▶ • Realm of ──▶ ...
  laws   biological evolution   cultural evolution

FIGURE 18.1

Alphakits and evolutionary dynamics shown at key levels of the grand sequence. The realm-starting partnerships of alphakit and evolutionary dynamics are at the base levels of the two evolutionary realms. The atomic alphakit as atoms–molecules occurs as a result of development that started with the fundamental quanta on the far left, with its "weak alphakit" shown in lighter gray. Not shown are the themes' other manifestations, both broader and finer-grained (see figures 17.3 and 17.4). The levels are: QUA (fundamental quanta), PRO (nucleons: protons, neutrons), NCL (atomic nuclei), ATM (atoms), MOL (molecules), PCL (prokaryotic cells), ECL (eukaryotic cells), ANI (multicellular organisms: animals), ASG (animal social groups), HUM (human tribal metagroups), AGR (agrovillages), STA (geopolitical states).

Right off, one parallel stands out: *both second levels were created by the integration of different prior things.*

- *Eukaryotic cells*: from the combination and integration of two distinct kinds of prokaryotes, bacteria and archaea.

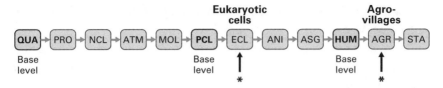

Dynamical realms:

- Realm of physical laws ────────▶ • Realm of biological evolution ────────▶ • Realm of cultural evolution ──▶ ...

FIGURE 18.2

The levels of the eukaryotic cells and agrovillages (*shown by arrows and asterisks*) *followed* the crucial base levels of the two evolutionary realms. In this chapter, we inquire into their parallels. The levels are as follows: QUA (fundamental quanta), PRO (nucleons: protons, neutrons), NCL (atomic nuclei), ATM (atoms), MOL (molecules), PCL (prokaryotic cells), ECL (eukaryotic cells), ANI (multicellular organisms: animals), ASG (animal social groups), HUM (human tribal metagroups), AGR (agrovillages), STA (geopolitical states).

- *Agrovillages*: from the combination and integration of earlier tribal metagroups (some of which had "grown" into complex hunter-gatherer societies) with plants and animals, which became domesticated and thereafter "inside."

Now, the eukaryotic cell merged participants from the same previous level. In contrast, the agrovillages directly merged two or more previous levels (tribal metagroups with multicellular plants and animals from two levels down). But this common pattern of merging of different things—right from the start of the event—is intriguing.

Contrast that pattern to the events of evolutionary combogenesis that merged things that initially were essentially alike.[1] Multicellular organisms involved the initial merging of cells of the same species; animal herds involved the merging of animals of the same species; tribal metagroups merged bands not inherently different; and geopolitical states merged agrovillages or their networks or chiefdoms that were not inherently different. About the state: the main distinct prior units were those that started exploring an expandable bureaucracy ahead of the others, but this was arguably the very defining innovation of the geopolitical state, not a preexisting, inherent difference.

There is more: *in both eukaryotic cells and agrovillages, one partner within the merger became a specialized organ involved with the control of energy in the new larger, individual.*

- The endosymbiotic bacteria in the merger into eukaryotic cells became the mitochondria, the new cell's power plants.
- Crops and domesticated animals gave human groups sources of energy (mainly food) within the new agrovillage networks. These domesticates were ultimately coaxed to give yields with increasingly higher efficiency and production.

This point about the parallel can be further refined: *as the merger progressed in each case, control was transferred.*

- In the creation of the eukaryotic cell, most genes of the endosymbionts went into the host's central genome (or were scuttled). This solved the potential "nightmare" of genetic conflicts of interests across scales, a solution necessary for integration.
- In agriculture, plants and animals kept their genomes, of course. But their genomes were altered as the plants and animals became tamed, integrated, and selected for desirable traits, such as larger grains that adhered to the seed clusters and more docile animals.

Admittedly, with the formation of every level subsequent to the origin of life at the prokaryotic cell, challenges involving control and coordination within the larger thing as well as challenges involving supplies had to be addressed because cells need imports of nutrients and energy. That is true for every level up to and including geopolitical states. (Grocery trucks cross the nation on state-regulated highways to feed cells in your body.) But what we have in the proposed parallel is unique: differentiation at the start of the union and one side becoming a specialized energy organ in the union! (I make only a passing reference here to dogs as emotional pals; anyway, dogs perhaps originated as workers for protection or on the hunt.)

Given these shared features, I suggest these two second, "sophomore" levels of the evolutionary realms are parallel, to some degree. So is there any common pattern to their origins?

It's possible that the innovations of the eukaryotic cell and agrovillage metagroup were parallel not merely in their independent advance in energy control enabled by specializing one partner of a merger. It's possible their innovations overcame a real limiting, "structural constraint" in both their respective, prior levels, a constraint involving energy and the inability to reach advantages:

- Nick Lane and William Martin's thesis (see chapter 9) is that an energy constraint limited the prokaryotic cell from doing what the eukaryotic cell was eventually able to do: evolve the forms of complex multicellularity to which we belong.
- Agriculture was able to overcome ecological, energy constraints that limited local band size of hunter-gatherers (as discussed in chapter 13; and even at the sites of resource superabundance the larger local populations were limited to being pinned to those sites).

Consider now that useful imaginary construct called a "possibility space."[2] Once the base levels of prokaryote cells and human tribal metagroups came into existence, there existed the *possibilities* of complex multicellularity and geopolitical states. Both possibilities offered advantages of size and complexity to further evolved systems that would have to stem, respectively, from the evolutionary base levels.

But the prokaryotes and tribes couldn't reach those advantages with the capabilities they had. What each could do and what each did do was participate in a next event of combogenesis that gave the new, larger things specialized "organs" of energy. That movement led to eukaryotic cells and agrovillages as second levels in their realms. Only from the latter levels were the next ones—complex multicellularity and geopolitical states, respectively—reachable. It's possible that in these crucial "second"-level cases, not only did the merger of unlikes provide an energy component, but such an energy component also overcame some strong degree of energy-related constraining factors of the prior, base levels. That would be an interesting deepening of the parallel between these second levels of the evolutionary realms of biology and culture.

Of course, the full complement of parts in this portrait of parallels rests on Lane and Martin's thesis regarding the eukaryotic cell being correct in terms of the prior, prokaryotic cell's energetic constraint. But the status of that part of the portrait does not affect the other, clearly delineated parallels in which both eukaryotic cell and agrovillage merged unlikes *and* gained partners as energy "organs."

Did this particular means of achieving an innovation in the control of energy—via an initial partner in the combogenesis—have to be at position two in each evolutionary realm? That would be difficult to claim. But it does seem to be what happened in the grand sequence here on Earth. Getting an energy revolution on board right after each base level makes some sense, at least in hindsight.

THE GRAND THEMES OF COMBOGENESIS:
MYSTERIES IN THE BASE LEVELS

One point struck me after having spent much time investigating the levels—how they came to be and what they led to. The origins of levels tend to involve some of the largest enigmas to our understanding. (I mostly but not exclusively refer to the levels in the evolutionary realms.) For example: the origins of the eukaryotic cell, animals, agriculture, geopolitical states. These enigmas are fields of research unto themselves, fields of waving flowers announcing question upon question from all their petals glistening in the sun of reality. One can stand before these fields in rapture. Entering them is even better.

Furthermore, the larger the transition between one level and the next, the deeper the mysteries tend to be. Where are the biggest mysteries? They're smack in the base levels. Consider the following.

The base level of the realm of physical laws, of the grand sequence, and therefore, basically, of the universe: the Standard Model and its fundamental quanta, such as quarks, electrons, gluons. The Standard Model, as a menu of potential across a profound array of mathematically described entities that form an integrated "set," was born all of a piece with the Big

Bang origin of our universe from a cosmic singularity.[3] Recall how fine-tuned the model is—for example, how the "sum of the masses of the types of quarks that make up a proton . . . seem roughly optimized for the existence of the largest number of stable nuclei."[4] In this way, this base level seems a lucky break for us, one that deserves our wonder and thanks for how it works and that there is something rather than nothing.[5] Some cosmologists do propose an ongoing cosmic evolution in that multiple lineages of universes get selected toward those of ever greater stability.[6] For now, for us, the Standard Model is what we have. Singular. Full of deep enigma.

Next, the base level of the realm of biological evolution: the prokaryotic cells and the origin of life. We know Eörs Szathmáry has stated that our understanding has recently progressed from "unknown unknowns" to "known unknowns."[7] What an expansive field of questions beckons to us—voices from LUCA, the last universal common ancestor. The origin of life and thus of all living cells is thought to have been singular. Evidence includes the universal genetic code, the universal set of amino acids as elements of an alphakit, universal phosphorus-based energy molecules, and many other universals across all living things. We don't yet understand enough about the intermediate levels that might have gone into the event, and thus we have pressing questions that we hope to answer by sending probes to Mars and to various moons of Jupiter and Saturn, which have water and other potentially favorable properties. Were there multiple, independent origins of life right here in our solar system?

Next, the base level of tribal metagroups and the origin of cumulative cultural evolution. *Homo sapiens* is the sole species on Earth today with cumulative cultural evolution and a linguistic alphakit. If the Neanderthals had not gone extinct, would they be flying in planes today? We know they passed on some genes to most of us. And there are questions regarding earlier species that walked upright. Wasn't the eventual evolution of cumulative culture and complex language already "set" by other distant relatives, such as the genus *Australopithecus*? The answer to this question is not at all clear. Furthermore, we saw that anthropologists are searching for an understanding that describes a primal structure of human sociality, trying out words with the prefixes *meta-*, *mega-*, *hyper-*, *ultra-*, *super-*, and so the questions are not just about brain size or tools. The use of "we" on multiple scales needs to be grappled with to understand this base level.

Thus, it appears that levels that start new dynamical realms have great degrees of mystery. Findings about these starts instantly make news. These base levels have so much to do with who we are and how we got here. The magnitude of their mysteries is likely related to the fact that these levels made particularly large innovations in their dynamics, which are difficult to unravel. So we have to understand how these base levels came to be.

THE GRAND THEMES OF COMBOGENESIS ACROSS THE GRAND SEQUENCE

The approach in this book has been to look at major themes across the levels in a comparative manner. That approach applies best to the two evolutionary base levels because both are inside the grand sequence, and we therefore can ponder both the before and the after of these levels as part of the sequence. And we have seen that the base levels of the biological and cultural evolution realms were associated with particular innovations that I have articulated as themes.

Those themes are evolutionary dynamics and alphakits. They came into play during the events that created life and cumulative culture. Figure 18.1 shows the locations where the themes appeared. The two themes helped each other, so to speak, in what I have called the partnership between alphakits and evolutionary dynamics.

Furthermore, inquiring into the formation of the two evolutionary base levels, we found it likely to assume evolutionary dynamics occurred in a kind of chemical evolution that preceded the start of life itself (for reasons I have explained, I place those chemical evolutionary dynamics at the level of molecules). We also found cognitive evolutionary dynamics in animals that were the kinds in our ancestry. Those cognitive dynamics carried over and were highly evolved into decision-making processes in human evolution. Thus, we were led to propose an anatomy to cultural evolution as dual-scale evolutionary dynamics: cognitive and social.

Key to all was the atomic alphakit of atoms and molecules as the two sets, which was inevitable from the Standard Model.

How might we conceptualize these patterns, themes, parallels? We saw that other kinds of parallels can be made, too—for example, in the similar means by which eukaryotic cells and agrovillages presumably were able to overcome energy-related constraints in their respective preceding levels.

Ah, overcoming constraints, limitations, blockages! Despite changes of things within a level, there might be opportunities in "possibility space" that would provide new sorts of stability and existence, but the things and relations are unable to reach those new lands if they stay within their level. Their jostlings can lead to combogenesis, however, and sometimes that process of combogenesis successfully moves the things on to new lands of opportunities through merger into new, larger things with innovative relations.

CONVERGENT COMBOGENESIS

Each transition to a new level opens up further new regions in the overall possibility space. This process is in line with my understanding of the complexity theorist Stuart Kauffmann's notion of the "adjacent possible."[8] Jonas Salk developed a similar concept using what he called "nonmanifest order."[9] The nonmanifest becomes manifest. We should somehow think about and take as real the nonmanifest, the adjacent possible, the metaphysical things of possibility space; we should at least think about major themes in the grand sequence, I suggest, as "creatures" of possibility that came into existence multiple times.

We can use the phrase *field of possibilities* to help ourselves understand the theme of the alphakit. The concept of nonmanifest possibility (to make sure all bases are covered) can be applied to the entire grand sequence. Yes, everything in the sequence had, in a sense, latent potential or possibility, from the start of the sequence with the matter-field quanta and force-field quanta of the Standard Model, that enigmatic base level of the entire sequence. But the application of such a conceptual possibility space or field comes into focus a bit better if we consider each level's things as capable of exploring a *subset* of the total space of possibilities, a subset that is restricted and special to each level.

So all our themes—evolutionary dynamics, alphakits, specialized energy subsystems from combogenesis—can be thought of as existing as possibilities, as nonmanifest order that becomes manifest, in certain cases as ways systems can take advantage of opportunities for existence. In the realm of physical laws, opportunity rests in the general concept of energy repose. In biological evolution, we have John Tyler Bonner's concept that the "top of the scale is always an open ecological niche," so that going beyond the current top (with its different meanings at each level) is a place in possibility space to seek for new ways of making a living. We saw the biological theme of conflicts of interest across scales, which had to be solved in each case.

Opportunities and challenges about size and complexity in the cultural realm were similar to those in the biological realm. The dynamics for achieving solutions to these challenges were special in the cultural realm, though, with the linguistic alphakit and dual-scale cognitive-social evolutionary dynamics.

Themes occur with local distinctions as the possible becomes real in various contexts along the grand sequence.

■ ■ ■

A powerful finding of evolutionary biology is embodied in the concept of convergent evolution. The wings of birds, bats, and butterflies share the property of being flat, which helps them fly. Furthermore, all of these delightful creatures evolved wings independently of each other. One might say their ancestors converged on the discovery of similar shapes for the function of flight.

The common bundle of features in the alphakit–evolutionary dynamics partnership is also a kind of convergence. Might we call the dual origin (biological, cultural) of this partnership an example of *convergent combogenesis*? To think about this possible phenomenon, we can view the alphakit, an abstract general pattern, as something like a mathematical function or the geometric properties of a sphere. The alphakit led to powerful consequences for the special things that first had new versions of it inside them: prokaryotic cells in the case of the genetic alphakit and tribal metagroups in the case of the linguistic alphakit.

And we might think of opportunities as a common pattern. Yes, the opportunities that the animal had (say, to evolve and to use eyes and ears) were different from the opportunities that the prokaryotic cells had. But opportunity itself is a theme, a kind of combogenic convergence, a theme with distinctions achieved at the different levels as they reached the embodiment of previous opportunities and opened themselves to new, unmanifest possibilities.

One further example of the broadly extended concept of convergence (not standard biological convergence!) is the existence and formation of "things and relations." This pattern is difficult to grasp because it "lives" at every level and therefore cannot be argued as having utility at one particular level more than at others. I have struggled with the best terms here and, as you have seen, have settled on *things* and *relations*. What do these terms refer to? Does "thingness" exist in possibility space, in the zone of nonmanifest order?

I am not going to get myself into more philosophical muddles than I'm already in. Throughout this book, I have taken a pragmatic approach: Are experts in various fields distinguishing "things" and "relations" that they have found useful and essential in their research? If yes, then those things and relations are good enough for me to use in this synthesis. The point about convergent combogenesis here is that the general pattern of things and relations appears multiple times in a variety of realizations throughout the levels of the grand sequence.

Combogenesis is itself clearly (I hope)—taking account of what I have said about opportunities and blockages, things and relations—a theme within the concept of convergent combogenesis. Combogenesis repeats. It is a "concept" in the possibility spaces of nonmanifest order that can sometimes and has, fortunately for us, come true.

For me, exploring these topics has led to an ever-growing awe of the cosmos for having things and relations on so many scales. Those flowers, beckoning! I truly think there is much more to do along these lines of thinking, and many people out there are already doing it in their own skillful, valuable ways. I have given you my approach in this book.

Starting with the simplest things of physics at the birth of our universe, a rhythm of combining and integrating led in a grand sequence of creation to living things and ever outward in growing scales of nestedness to the first civilizations. Each level or scale created larger and ever new things, each with powerful new relations. Twice within the overall family of levels—once at the start of life and once at the start of culture—the newly formed things contained alphabet-like innovations and launched a special style of evolutionary dynamics with themes in the previous and subsequent levels. The process might remind one of musical themes that hit similar but higher notes as they ascend up in octaves.

May we enjoy this amazing world we are lucky to be in, and may our knowledge help us create desirable future levels of the grand sequence, should they come.

EPILOGUE

What About the Future?

SUMMARY: We can use aspects of the levels of the grand sequence as ways to frame and explore questions about our future. For example, are we today still within the level of the geopolitical state? Might we be on a ramp of combogenesis from this level toward the next or even already on the cusp of the new one? At minimum, one would expect that at the next level the prior innovation of the geopolitical state, with its capability for unlimited acquisition and merger, would get reconfigured internally into stable relations of nations within the new, larger thing. To some degree, that does seem to be happening. But we would expect more radical changes, too. Indeed, can the new level even initiate a new dynamical realm? These questions lead us into considerations about individuals and the planetary scale.

AFTER THOUSANDS OF YEARS IN THE GEOPOLITICAL STATE

We all face the big unknown of humanity's future. So much about the world seems in transition. Where are we headed? Can the levels I have described be used as guides for what might be coming? Here I offer some preliminary exploration of these questions, using patterns and principles developed in this book. I suggest that the concepts of combogenesis and the grand sequence can at the very least provide a viewpoint.

First, it seems to me that for thousands of years, at least until fairly recently, we have been in the level of the geopolitical state. That level was reached, as described in chapter 14, at various dates at a number of primary sites around the world. The main innovation was the expandable hierarchy of bureaucracy and the ability to grow through acquisition and merger.

The cultural modifications and applications of such hierarchies have been the mainstays of politics and thus of human life for the millennia since the state started. Evidence of recent modes of growth (realized or attempted) by takeover are still fresh in the collective memory from the two world wars of the twentieth century. We might include in the general pattern the behaviors of corporations (as statelike entities) with assimilation at various scales. Charles Spencer, whom we heard from on the origin of geopolitical states and territorial expansion, points to deep continuity from ancient times to today's world: "When the bureaucratic decision-making design is adapted by any organization—whether governmental, for-profit, or non-profit entity—we should not be surprised if it engages in predatory behavior."[1]

Huge changes have clearly happened since the development of the ancient states at Uruk, Monte Albán, and Mohenjo-daro. Various forms of state organization have included feudalism, republicanism, and modern democracies. There has been intense cultural evolution across the styles of organizations noted. The printing press is often cited as an invention that transformed the flow and network of information. But, as I hoped to have shown, in the grand sequence we can distinguish between changes (even large ones) *within* levels (worms becoming whales within the level of multicellularity) and transitions *between* levels (whales themselves as members of a social group on a new level). To my reading, for thousands of years the changes in large-scale political organizations have been *within* the state.

And yet so much is in flux. James Lovelock, originator of the Gaia hypothesis, cites the Industrial Revolution as the start of what many are calling the *Anthropocene*, a relatively new, spreading term for a geological epoch in which human influence affects the entire planet.[2]

The takeover mode of the geopolitical state has to a large degree, though not totally, fallen out of favor internationally. We are witnessing experiments with supranational structures such as the European Union, multination trade agreements, as well as the globe-spanning treaty to protect stratospheric ozone and efforts to coordinate reductions of the CO_2 emissions that are accumulating in the atmosphere.

Buzzwords evoking a transition include *globalization, Internet,* and *instant news cycle*; to that list we can add what the biologist Edward O. Wilson calls the "technoscientific global community of today."[3]

In the spirit that something is happening—a metamorphosis of some sort—we might ask about a new level of the grand sequence. Are we on its cusp? Perhaps we are not yet on that new level, or perhaps the event has just begun, or maybe it is already well along in process.

A NEW LEVEL?

Following the logic of combogenesis, we would expect that a new level would build from things, systems, entities, ontums of the prior level. So as a first crack at characterization, we might say that this next level will be made from component geopolitical states.

But Charles Spencer has noted that just getting larger in the case of the state does not mean something truly different. Thus, we would need more than a growth in size to gain confidence in pointing to a new level. We would require the new entity to possess some sort of truly new relations. It seems to me such new relations could develop when the suprastate "thing" is global. At the planetary scale, there is nothing more to expand into. Let's hold off for a moment on the issue of outer space.

I first gained the concept of a planetary scale of humanity from the writings of Teilhard de Chardin (1881–1955).[4] At the planetary scale, an expanding phenomenon—geochemical, biological, cultural—closes back upon itself. Teilhard (as he is often called) noted that when culture becomes planetary, it runs into itself. It starts folding back on itself, which can trigger the birth of a new order. For our times, Teilhard called the new order the "noosphere," the sphere of *noos*, the Greek word for "mind." Some say Teilhard (and others) anticipated the Internet.

For Buckminster Fuller, inventor of the geodesic dome and famed user of the term *Spaceship Earth*, a planetary milestone was reached with the British Empire because it was a "spherically-closed, finite system."[5]

But even with the British Empire, takeover was still possible in unconquered or uncolonized lands. At a future, truly global scale, however, the new thing to be formed cannot operate in takeover mode toward the outside because all peoples and lands are already on the inside. And

I surmise that getting to that new system in a stable way will not involve a takeover mode, the bureaucratically driven predation that Spencer notes. If in the "new" system one nation or small group rules, that doesn't really seem new. That would not be a new level, except for the fact that there would be nothing more to take over afterward. Or perhaps the mode of conquest would move on into space.

Even if globalization to a new level does not happen by takeover and so at least hints at something innovative in the wings, that does not mean that takeover is not operating at all internally—as adjustments, shall we say. But this mode (or, better, these modes of variation within a type) will likely alter significantly if what happens is a distinct new level. The issue of conflicts of interest across scales, which we saw in events after the origin of life (for all levels after the start of biological evolution and continuing into cultural evolution), would have to be solved to reach this new level.

Living cells, for example, changed significantly when they incorporated into multicellular bodies, a transition that happened multiple times and that led to three familiar lineages: fungi, plants, and animals. At the new level of complex multicellularity, cells inside these bodies were still cells, needing to import nutrients and energy and to export wastes (now to within the larger body), but they cooperated in an evolved division of biochemical labors and other wonders.

Given these considerations, we don't seem to be at the new level yet. I don't think the transformations implied by solving the cross-level conflicts of interests are yet here. Many of those people cheering the Internet think of it as a disruptive, empowering tool that will effect some sort of radical change, which we perhaps barely as yet glimpse. Science fiction and Internet philosophy explore these possibilities.[6] The general phenomena noted earlier do make me think we are somewhere on the ramp of a combogenesis event. Early or late on it? To find it, we must look for major changes in nations and even in people. We must look for changes that are likely *more radical than can easily be imagined*. The end or at least transformation of presidencies and prime ministers as institutions, for instance.

Jonas Salk, who invented the vaccine for polio and was noted in this book's preface, was engaged toward the end of his life in thinking big about our planetary human future. He wrote that we need a transition to "conscious creative metabiological evolution as a process by which the future of humankind will be determined."[7] Perhaps conditions of life seem all right for now (for many in developed nations), but we will likely require significant innovations in the supranational structures hinted at earlier if we are not merely to muddle ahead in uncertainty, given the need to ramp up gross world product by perhaps threefold for overall prosperity and to meet environmental challenges. Here is how Albert Einstein, writing in 1948, put it: "We must revolutionize our thinking, revolutionize our actions, and we must have the courage to revolutionize relations among the nations of the world."[8]

THE ROLE OF THE INDIVIDUAL

Individuals are several levels down from this hypothesized new level. To review, we humans *live on and are* several levels of existence. As culture-imbued minds, we are parts of the webs of "we" at the tribal metagroup level (transformed over thousands of years—for example, in the university where I teach, with its Paleolithic-band-size departments). We are beings within animal social groups, such as extended families at the next deeper level, and our bodies are animals on a level prior to that. A crucial aspect of who we are is our capacity for cognitive evolutionary dynamics to roll along daily in our heads and minds—minds that through biological evolution came "up" from primitive form to become one of the pair of sophisticated contributors to the overall dynamics of cultural evolution (the other is that of the more recent, social evolutionary dynamics.) An outstanding question is what might happen to our individual, cognitive evolutionary dynamics at a level beyond the geopolitical state.

The Borg are one of the alien races in the universe of the *Star Trek* series. They can incorporate other races by turning individuals into cyborg beings, combinations of the biological and the electronic, all in service of the hive, which is run by a Borg queen. Great stuff! At least to watch, that is. This is one possible pathway to a bizarre and frightening (to me, at least) incorporation into a greater whole, which, as in the Borg case, may span multiple star systems. We will have to beware borgification (if I may) in any event of combination and integration to some emerging new level. We might want to preserve our integrity as our inner dynamics become, we hope, even more enhanced and wonderful.

What will that enhancement mean? Enlightenment? The sound of one hand clapping and inner peace as we enter a vigorous, complex oneness that is social as well? Others are concerned about the possible directions of this "enhanced" future, too. Edward O. Wilson hopes we do not try and compete with "robot technology by using brain implants and genetically improved intelligence and social behavior," and he promotes "the preservation of biological human nature as a sacred trust."[9] James Lovelock is not nearly as concerned about a merger of biology and electronics and even makes an analogy to the ancient biological event of endosymbiosis (discussed in chapter 9), in which different species of prokaryotic cells fused into the eukaryotic cell.[10] There might be a major conflict between members of the growing transhuman movement and those who support the preservation of human nature. To some degree, *The Matrix* film series (Wachowski Brothers, 1999, 2003) was about such a conflict, though current transhumanists would want the new humans to play a more significant role than as batteries in pods fed by tubes.

The world currently is complicated. At times, it is a stressful mess. A radical planetization would have to include a new, larger scale of the social evolutionary dynamics we already have but also incorporate the internal, cognitive dynamics of individuals. That is the general guideline from combogenesis. But what could that mean in realization? Who we are is partly who we feel ourselves to be during our personal decision making, and I hope it will be the case that our choices will contribute to any new system of a next level. What kind of service will individuals perform? Will there be a new ethics?[11] The individual has to participate in such changes if human nature is to be preserved.

But what kind of participation? Such questions need to be asked and perhaps addressed fairly soon, or they might be answered for us by trends larger than we can see.

Jacob Bronowski, noted in the preface for his concept of stratified stability, a forerunner to this book's concept of combogenesis, championed "a democracy of the intellect."[12] Here is Jonas Salk again, weighing in on this issue: "Individuals can bring about change more easily than can institutions. Therefore individuals are of particular importance, since they are more flexible and adaptable to changing circumstances."[13] And Einstein in 1931: "I regard it is the chief duty of the state to protect the individual and give him [or her] the opportunity to develop into a creative personality."[14]

From the perspective of the grand sequence, to ground the continued cultural evolution of the individual self, we should celebrate the level of the animal social group. In the type of group from which our ancestors evolved, animals were physically separate, brainy individuals. Not able to physically merge like cells of our bodies or to join in contact like polyps in a coral colony, our animal ancestors were connected in groups via their senses that could foster fluid relationships across space, a precondition for language, which later was to ride upon modulated sound waves at the subsequent level of the human tribal metagroup.

NEW RELATIONS OF A HUMANIZED BIOSPHERE

Will the global level be necessary to ensure the fulfillment of what Einstein saw as the chief duty of the state? It might be, given the need to avoid possible predatory behavior of bureaucratic decision making and yet continue with ongoing progress to global prosperity, which is putting pressure on many institutions. By 1949, the fact of the big scale was clear to Einstein: "It is only a slight exaggeration to say that mankind constitutes even now a planetary community of production and consumption."[15] So our current organizations are evolving within this great community and the third-rock-from-the-sun reality of all surface systems folding back upon themselves. There is simply, for now, nowhere else to go within the finite film of Earth's biosphere, now filling up with cultural metagroups. This humanized biosphere includes all people and everything they have brought into their systems via tens of thousands of years of micro-combogenesis and the major transitions of combogenesis that have taken place.

But then what about new relations of the planetary community? In this book, new relations of new things imply hookups with other things outside: atoms able to connect into molecules, prokaryotic cells able to connect into eukaryotic cells, animals able to connect into social groups. But for us the outside cosmos is empty space, stars, and radiation. Is that fact of a barely scratchable (so far) physical cosmos at the heart of the new relations?

We living in the biosphere also have relations to the deep earth, to fossil fuels, to minerals, to groundwater. And thus the humanized biosphere as a whole has those relations. We are forging a new relationship to the sun in the form of solar energy. Can we talk about advantages to the new thing, assuming it is starting to coalesce? Such advantages were important at previous evolutionary levels, and a thing at one of those new levels needed to provide advantages to the included, former levels as well. Given the physical cosmos way outside and the rocks way down inside Earth, we might have to admit that the new, odd relations might be the very absence of external "like" systems. (As far as we know, Earth has no competition yet, despite plenty of "contact" in science fiction.)

However, let us also consider the biosphere in people's minds. We as individuals might envision ourselves organizing at the planetary scale (it is the reality, in any case). If we can discuss the global scale, we have a relationship to it, and it to each of us.

WHAT ABOUT A NEW DYNAMICAL REALM?

As we have seen in the grand sequence, sometimes a new level starts a new realm. Could a level we might be approaching (or already be in) be the base level of a new dynamical realm?

Given that culture is an evolutionary realm that came from a prior evolutionary realm of biology, a new realm after cultural evolution would presumably also be an evolutionary realm. As we have seen, a partnership of innovations comes in at the base level of an evolutionary realm: the starts of both biology and culture meant a new alphakit and a new style of evolutionary dynamics with propagation, variation, and selection.

Might computer code be a new alphakit? At its most reduced, computer code is binary: the "bit" is either on or off, present or absent. One might say it is simply another example of the multiple, alphabet-like systems the human mind has invented, including not only all-important language but also music and math, all of which have alphakit structures. But perhaps computer code is newly different enough to pay special attention to because its operations can exist externally to the body, exosomatically. The codes and structures built from language, music, and math obviously exist exosomatically, too, most obviously in writing, but their operations are still inside their creators' heads or get transferred to and exchanged among other people's heads. The growing interdependence of more and more sophisticated "metabolisms" of multiple scales of computer code fuels some of the concerns about cyborg people and about artificial intelligence running independently from people.[16] Science fiction is of course hugely involved in exploring this issue—working ahead of its time, titillating, stimulating, provoking, and warning.[17]

What about the dynamics of a coming evolutionary realm? Well, there is conscious use of evolutionary principles in the production of designs: in robotics, in artificial intelligence, in drug design, and in much more.[18] The concept is to build in variability as a way of exploring the possibility space for some class of systems and then evaluating the results, the selection strand in the trio of strands of evolutionary dynamics. Some of the productions go into the next round of propagation and variation. A best solution can then be disseminated. Might the use of such dynamics, which can be set running automatically without needing humans—thus way beyond Thomas Edison's trying out of different filaments for an electric lightbulb—be signs that the entire organized planet will be the base level of a new evolutionary realm?

But wait. Are the new types of evolutionary dynamics used in computer science, engineering, and general design simply augmented components within the already-existing social evolutionary dynamics of complex group decision making among people, which have existed since the start of cultural evolution? How might the new types be part of an event to a new level that gives birth to an entirely new evolutionary realm? Given that in the grand sequence the graduation, final levels of physical law and biological evolution contained precursors to the new alphakits and dynamics of the subsequent biological and cultural evolution realms, respectively, what we are seeing today in computer code and evolutionary algorithms for designs at various scales might be precursors to something much more truly radical when (and if) the next level forms. Depending on how one looks at these various trends, the emerging planetary level might be the final level of the realm of cultural evolution, just before the next level that really rockets off into something likely barely imaginable to us.

A new evolutionary realm could be so transformative that we can scarcely envision its possible shape and operations from our current viewpoint.

■ ■ ■

In general, an integrated system of geopolitical states is the easiest result to predict as a next level in the logic of the grand sequence. We see hints and experiments occurring along these lines. But, please, if this means a world government that is just larger than the ones that currently exist, and if this means that certain individuals end up with even more power than the powerful currently possess, I emphatically would not consider this integrated system a new level on the grand sequence. There are questions to be considered about what happens to all things of the preceding levels as the new planetary level develops. For example, until recently most people on Earth had some direct connection to a participation in agriculture. With fossil fuels, that is no longer the case. We can expect changes in all the levels as they become nested more deeply inside one more new level being formed by combogenesis, if that indeed is what is taking place.

I have highlighted some aspects of current trends and challenges in the modern world that seem relevant from a viewpoint of concepts developed from the grand sequence. It is not clear how much the whole of a purported new level or realm is visualizable and understandable inside our minds because the scales are getting huge. This challenge for our minds is happening already today. But, to reemphasize a point I made earlier, I suggest we try and solidify our current cognitive dynamic as cool, joyful, fulfilling participants in the larger evolving system. As we head into the future with its possible new event of combogenesis, let us not have our cognition subsumed and dominated by complex nestings of some sort of social-electronic superevolutionary dynamics. Let us actively sculpt a concept of personal enlightenment and refine that concept as the future progresses.

ACKNOWLEDGMENTS

First, thanks to scholars who gave their time and accepted my request to provide technical reads of the chapters that cover their specialties. Many of these experts are cited by publication and/or personal communication in this book. The revisions I made after their reviews were aided by their comments, and I trust I did not make substantial changes that they would not agree with. Of course, all errors are my responsibility, and I will endeavor to make corrections in the future. Therefore, I thank

Allen Mincer, professor of physics, New York University (chapters 3 and 4: fundamental quanta, nucleons).

Richard Orr, clinical assistant professor of chemistry, New York University (chapter 5: atomic nuclei).

Anna Powers, Ph.D. in theoretical chemistry, CEO and founder of Powers Education (chapters 6 and 7: atoms, molecules).

Addy Pross, emeritus professor of chemistry, Ben Gurion University of the Negev; visiting professor at New York University Shanghai (chapter 8: prokaryotic cells and the origin of life).

David Schwartzman, professor emeritus, Department of Biology, Howard University (chapters 8 and 9: prokaryotic cells and the origin of life, eukaryotic cells).

Neil Blackstone, professor of biology, Northern Illinois University (chapter 9: eukaryotic cells).

Eric Brenner, clinical assistant professor of biology, New York University (chapters 10 and 11: multicellular organisms, animal social groups).

Andrew Bourke, professor of evolutionary biology, University of East An-
glia, (chapter 11: animal social groups).

Anne Rademacher, associate professor of anthropology and environmental studies, New York University (chapter 12: tribal metagroups and the
origin of culture).

Kim Hill, professor of anthropology, Institute of Human Origins, School
of Human Evolution and Social Change, Arizona State University (chapter 12: tribal metagroups and the origin of culture).

John Hoffecker, fellow, Institute of Arctic and Alpine Research, Boulder,
Colorado (chapter 12: tribal metagroups and the origin of culture).

Rachel Meyer, executive director of the Conservation Genomics Consor-
tium, University of California at Los Angeles (chapter 13: agrovillages).

Charles Spencer, curator of Mexican and Central American archaeology,
Division of Anthropology, American Museum of Natural History, New
York; professor, Richard Gilder Graduate School (chapter 14: geopoliti-
cal states).

I also thank the initial anonymous reviewers who accepted requests
from Columbia University Press to respond to the proposal, which included draft chapters, as well as the anonymous reviewers who accepted
the press's request to review the entire draft manuscript. Their comments
helped me more than they will ever know.

I am indebted to and thank New York University faculty who attended
my faculty seminars to test this material, including those in the Departments of Biology and Environmental Studies; the many bright and engaging students who took my Advanced Honors Seminars at New York
University, "Metapatterns from Quarks to Culture" and "Transdisciplinary
Investigations at Multiple Evolutionary Scales," with gratitude to Michael
Fischer for organizing them; Nathalie Gontier for great conversations
and the invitation to speak in the plenary at the Lisbon conference on
evolutionary patterns as well as colleagues there, including Matt Haber,
Daniel McShea, Emanuele Serrelli, Davide Vecchi, and Richard Watson
for tuning in well enough to give me enthusiastic suggestions; attendees
at my seminar on this material at Columbia University's sustainability-
development seminar, especially Eyal Frank for the invitation; Kathelin
Gray for conversation and for the invitation to present this material at

Synergia Ranch in New Mexico, which engendered useful comments from John Allen, Bill Dempster, and Mark Nelson; David Rothenberg for the invitation to present at a "Wonder Cabinet" (which included Ren Weschler); "meta-amigos" Jeff Bloom and Bruno Clarke for wide-ranging systems syntheses, with a special nod to Jeff for ongoing input across cycles of time and to Bruno for comments on the proposal; John Silbersack for hearing me out and for help shaping the proposal; Francesco Tubiello for big-picture comments about the entire draft manuscript; Peter Westbroek for reading the manuscript and for discussions; longtime Amygdaloids band mates Joseph LeDoux and Daniela Schiller for conversations way beyond music and relevant to this work; Connie Barlow, Jennifer Jacquet, and Colleen Silky for comments on a draft of the preface; colleagues across so many fields at New York University for systems conversations and expertise, such as Dale Jamieson and "Socrates" Bill Ruddick for points about language, Trace Jordan, Neville Kallenbach, Mary Killilea, Fred Myers, Michael Rampino, Christopher Schlottmann, Richard Wener, Marina Zurkow, and Dan Zwanziger; brother Ken Volk for comments on several chapters; Luke Wallin for overall enthusiasm and discussions about ontology and terms; Julianne Warren for systems talk about generativity; Marty Hoffert for wide-ranging insights; David Sloan Wilson for conversations; Sérgio Faria for his interest and friendship; Mary Evelyn Tucker for interest; Christie Henry for enthusiasm on the project; O. Roger Anderson, Bruce Bugbee, Susan Doll, Joe Franceschi, John Horgan, Lyn Hughes, Axel Kleidon, Seymour Reichlin, Laura Schwartz, and Mitchell Thomashow for relevant systems conversations; Jeff Greenberg and Sheldon Solomon for insights into life's conditions; all my friends and colleagues in systems thinking over the years for conversations about the biosphere, the global carbon cycle, NASA's advanced life-support program, metapatterns, evolution, and more; and, finally, my siblings for asking and asking. I apologize to anyone I have inadvertently left out.

Diane Dittrick of Barnard University and the New York University Tandon School of Engineering was enormously helpful in her review of a draft, wisely convincing me to eliminate confusing terminology and offering advice on much more too extensive to detail. Let me just say that when it came time for me to make a final revision, her thoroughly marked-up copy was by my side.

Amelia Amon commented on the earlier draft, and our talks motivated a crucial change in conceptual direction. She has been unceasingly supportive and helped with a brainstorm about the book's title when we were standing in line one morning to enter the Prado in Madrid. This book is dedicated to her.

Finally, I extend my gratitude to staff at Columbia University Press: Patrick Fitzgerald, editor and publisher for the life sciences, for his vision about this project; Ryan Groendyk, assistant editor, for making the process a pleasure; and the other individuals at the press who worked on publishing this book. It is great to be "back" at the press after a bit more than twenty years. I also thank Annie Barva for great copyediting and getting me through the crucial, final few percent.

GLOSSARY

ALPHAKIT: At minimum, a linked pair of sets—an element set and a cornucopia set.

ATOMIC ALPHAKIT: The alphakit with atoms as the element set and molecules as the cornucopia set.

BASE LEVEL: The first level of a dynamical realm, which establishes the dynamics that continue across the levels of that realm.

BIOLOGICAL EVOLUTION: In this work, the standard, high level of description of the processes that Charles Darwin described in the famous "final" paragraph of *On the Origin of Species*, here characterized (and generalized) in three terms: *propagation*, *variation*, and *selection*.

CHEMICAL EVOLUTION: An era theorized to have taken molecules to the origin of life by way of a progressive replicative chemistry that presumably had the three stands of a general evolutionary process.

COGNITIVE EVOLUTIONARY DYNAMICS: An evolutionarily early example of these dynamics is instrumental learning in animals, in which cognitive processes exhibit the trio of propagation, variation, and selection. Decision making inside one's mind also uses this trio but in very complex ways (yet still can be called cognitive evolutionary dynamics).

COMBOGENESIS: The genesis of new types of things by combination and integration of previously existing things, restricted in this book to the types along the levels of the grand sequence.

COMBOGENIC CONVERGENCE: The arrival at themes or parallels in the patterns produced by combogenesis along the grand sequence.

CORNUCOPIA SET: In an alphakit, the large number of realized composites of the members of an element set and within an even more numerous field of possibilities.

CULTURAL EVOLUTION: The culturally created evolutionary process that includes a dual-scale evolutionary dynamics: cognitive and social. It begins with the level of tribal metagroups.

DUAL-SCALE EVOLUTIONARY DYNAMICS. See **CULTURAL EVOLUTION**

DYNAMICAL REALM: A group of adjacent levels that share core dynamics.

ELEMENT SET: In an alphakit, the set that contains a small number of classes of discrete, relatively simple things from which the cornucopia set is produced.

ENTITY: Synonym for *thing, system, ontum* (see **THING** and the discussion in chapter 1).

EVOLUTION: Any general process that produces patterns from a braided, cyclic application of the trio of subprocesses or strands: propagation, variation, and selection (see discussion in chapter 17).

EVOLUTIONARY REALM: A dynamical realm of adjacent levels in which the core dynamics are (biologically or culturally) evolutionary, as defined in this book.

FIELD OF POSSIBILITIES: The totality of possible states that things in an alphakit's cornucopia set can attain (and might) but have not done so, for whatever reason.

GENETIC ALPHAKIT: The double alphakit that started with the origin of life, consisting of bases/codons as elements and genes as cornucopia coupled to amino acids as elements and proteins as cornucopia.

GRAND SEQUENCE: The series of levels built by the iterations or cycles of combogenesis, starting with the fundamental quanta of the Standard Model of particle physics and ending (in this work) with the geopolitical state.

LEVEL: A category or set of things or systems along the grand sequence (e.g., the level of atoms, the level of eukaryotic cells, and so forth). I sometimes use the term *level* to refer to an entity on that level.

LINGUISTIC ALPHAKIT. The alphakit that began with the level of cultural evolution, consisting of spoken phonemes (the elemental units of speech) as the element set and words as the cornucopia set.

ONTUM: Synonym for *entity, thing, system* (see the discussion in chapter 1).

PROPAGATION: One of the trio of basic processes in any kind of evolutionary dynamics (biological, cognitive, cultural). Sometimes called "reproduction" or "heredity" (in which case it also needs to be defined to apply to culture). It is the process that maintains the continuation of patterns.

REALM: Short for **DYNAMICAL REALM**.

RELATIONS: The properties of things that allow them to relate or interact with other things.

SELECTION: In the trio of basic processes in any kind of evolutionary dynamics, the process or processes by which some patterns in a field of variants are given preferential ability to propagate.

SOCIAL EVOLUTIONARY DYNAMICS: Dynamics among people that manifest by actions of individuals (such as market dynamics) or incorporate by social design (such as collective decision making) the trio of strands of a generalized evolutionary process (see **EVOLUTION**).

STRAND: See **EVOLUTION**.

SYSTEM: Synonym for *entity, thing, ontum* (see the discussion in chapter 1).

THING: Synonym for *entity, system, ontum* (see the discussion in chapter 1). Of the choices, the most commonly used term in this book.

VARIATION: In the trio of basic processes in any kind of evolutionary dynamics, the process or processes by which patterns are altered so that populations of those patterns contain differences, which can be subject to selection.

NOTES

PREFACE

1. Volk 1995. See also Volk and Bloom 2007; Volk, Bloom, and Richards 2007.
2. Bronowski 1973:348–349.
3. This distinction requires a way to define fundamental levels, which is an initial aim in this book. Thus, I distinguish between the building up from one level to the next (such as protons and neutrons becoming atomic nuclei), on the one hand, and the increase of diversity within a level, on the other (such as the diversification of atomic nuclei in stars or the evolution from simple to complex multicellular animals). Bronowski had a great insight and a major impact on me. He did nail the large difference in the behavior of things when evolution started, a theme in this book.
4. For approaches to and debates about the major transitions of evolution, see Smith and Szathmáry 1998; D. Wilson 2007; Bourke 2011; Calcott and Sterelny 2011; McShea and Simpson 2011; Szathmáry 2015 (for a "2.0" update); O'Malley and Powell 2016. For degree of nestedness, see McShea and Brandon 2010.
5. Salk 1983, 1985.
6. E. Wilson 1998.
7. For emergence and networks, see any of numerous works by Stuart Kauffmann and Albert-László Barabási (such as Kauffmann 2008 and Barabási 2003) and then follow the trail to those who discuss them. For synergy, see Corning and Szathmáry 2015.
8. For a definition of emergence, I like the simple directedness of the philosopher and systems thinker Mario Bunge: "Wholes possess properties that their parts lack. Such global properties are said to be *emergent*" (2003:12). There is so much written on "emergence" available by searching. Relevant books include Johnson 2001 and Morowitz 2002.
9. Some of this work is described in general terms in Volk [1998] 2003. Selected papers are available at http://metapatterns.wikidot.com/members:tylervolk. See also Volk 2008 and my vita at http://www.biology.as.nyu.edu/object/TylerVolk.html.

1. NATURAL CHAPTERS AND NESTED SCALES

1. See Volk 1995 for a cross-disciplinary discussion of arrows, break, stages, cycles.
2. Bianconi et al. 2013:463.
3. My estimate, using an average cell diameter of 50 microns.

2. THE CORE THEME: COMBOGENESIS

1. In a fascinating paper, Nathalie Gontier (2007) uses the concept "universal symbio-genesis" in a way very similar to my concept of combogenesis. I have explained to her, however, that for me the term *symbiogenesis* does not work when applied to physics and culture because of the *bio* part, a heritage from a word developed to apply to biological evolution.
2. Systems pioneer Ludwig von Bertalanffy's ([1968] 2015) use is in line with mine. Instead of *things* and *relations*, Robert Hazen (2009) uses the terms *agents* and *interactions*.
3. Hawking and Mlodinow 2010.
4. *Leviathan* is available as a free ebook from Project Gutenberg.
5. In Volk 1995, I discuss how the "center" of certain systems embodies (by control or other means) the dynamics of the whole system.

3. A BIG BANG START OF THINGS AND RELATIONS

1. Wilczek 2008:165. "It would be hard to exaggerate the scope, power, precision, and proven accuracy of the Core," states Wilczek (165). (*Core theory* is his preferred term for the Standard Model.)
2. Schumm 2004.
3. Quoted in Gribbin 2013:117. For other strong remarks on the deep mathematical perfection of this level, see Schumm 2004:3, 101–102, 139, and Wilczek 2008:33.
4. Tegmark 2014. The physicist Allen Mincer, a colleague at New York University, prefers to characterize the fundamental quanta as physical objects with mathematical properties (personal communication, September 2016).
5. Randall 2011:116. Randall elaborates, "Standard Model predictions work if we just assume particles have these masses. But we don't know where they came from in the first place" (116).
6. Veltman 2003:66, 306.
7. Quoted in Randall 2011:94–95.
8. Bruce Schumm cites this size limit and says quarks might even be "pointlike" (2004:186). And Martinus Veltman says that "for all we know they are point-like" (2003:12).

9. Randall 2011:116.
10. Allen Mincer, personal communications, February 2012 and September 2016. Also see Veltman 2003:71.
11. Close 2004:5.
12. Wittgenstein 1922:proposition 7, also at https://en.wikiquote.org/wiki/Ludwig_Wittgenstein.

4. THE NUCLEONS, WITH IMMORTAL PROTON AND FRAGILE NEUTRON

1. A quark–gluon plasma has been briefly created in experiments, but the ratchet here was nonreversing for all practical purposes.
2. Veltman 2003:296, 225.
3. Strassler n.d.b. And Strassler's definition of zillions (at the same website) is apropos: "too many and too changeable to count usefully."
4. Allen Mincer, personal communications, February 2012 and September 2016.
5. Schumm 2004:306.
6. Strassler n.d.a.
7. Schumm 2004:306.
8. Wilczek 2008:passim. Wilczek borrowed the phrase from the physicist John Wheeler (1911–2008). Andreas Kronfeld (2008) reports on calculations of the proton's mass using lattice-gauge theory.
9. Strassler n.d.a.
10. Schumm 2004:130, see also 94.
11. Close 2004:5.
12. Schumm 2004:133.
13. Vayenas and Souentie 2012:chap. 3.

5. ATOMIC NUCLEI FROM MUTUAL AID

1. The original masterful book about this event is *The First Three Minutes: A Modern View of the Origin of the Universe* (Weinberg 1977).
2. In the song "Woodstock," a great evocation for us all.
3. Hawking and Mlodinow 2010:160.
4. Reynolds 2015.

6. ATOMS WITH SPACE-FILLING, ELECTRIC MANDALAS

1. Close 2004:3.
2. Strassler n.d.a.

3. Physicists say that the tripleness of color relates to profound symmetry considerations. A great explanation of it is in Schumm 2004, involving Lie groups and conservation laws.
4. Jordan and Kallenbach have used this term in many conversations with me.

7. AN EXPANDING CORNUCOPIA OF MOLECULES

1. Lucretius 1994:book 5, ll. 422–426.
2. Lucretius 2010:passim.
3. Lucretius 1994:book 5, ll. 430–431.
4. Gregory Gabadadze, New York University physics professor, personal communication, December 2013.
5. Darling n.d.
6. Translator Ian Johnson says one of Lucretius's favorite analogies is "comparing the
   letters of the alphabet used in the formation of words with the primary particles used
   in the formation of substances. The analogy is all the more pertinent in Latin because
   the word *elementum* [plural *elementa*] refers to both letters and particles" (Lucretius
   2010:xiii n.)
7. Hazen et al. 2008:1693.
8. Falkowski 2015:95.
9. See Dryden, Thomson, and White 2008 and Riddihough 2015 for issues regarding
   counts and functionality.

8. SIMPLE CELLS LAUNCH LIFE AND EVOLUTION

1. Pross 2011, 2012, 2013; see also Pascal and Pross 2014. I also note here the concept of
   "autopoeisis."
2. Lane 2009.
3. Morowitz 2002.
4. Addy Pross, personal communication, May–June 2016.
5. Fossil evidence or chemical evidence by carbon isotopes typically indicates an even
   earlier date.
6. For example, see Smith and Szathmáry 1998 and Szathmáry 2015.
7. Pross 2011, 2012, 2013; Pascal and Pross 2016.
8. Szathmáry 2015.
9. Ibid.:10106.
10. These three processes can be found in the final paragraph of *On the Origin of Species*
    (Darwin 1859). David Sloan Wilson (2007) describes the "ingredients" for a three-part
    "recipe" of evolution.
11. Many researchers use the term *chemical evolution*—see, for instance, Martin 2011 and
    Pross 2011.

9. THE SEXY EUKARYOTIC CELL

1. Smith and Szathmáry 1998. Also see the sources cited in note 4 for the preface.

2. Woese and Fox 1977; Woese, Kandler, and Wheelis 1990. See Falkowski 2015 for a current, well-written general discussion.

3. The date is rough; it has quite a bit of "plus or minus" in it, depending on what evidence is being garnered and debated.

4. Margulis 1971. Lynn Margulis (1938–2011) was also a key contributor to the Gaia hypothesis, debates in which I was an active contributor but also a critic (Volk 2002, [1998] 2003). You can watch the first eleven minutes of my remarks on the second day of the "scientific" memorial for Margulis, March 24, 2012, at https://www.youtube .com/watch?v=urjZ286WFRw.

5. Falkowski 2015:116–123.

6. Gontier 2007; Blackstone 2013.

7. See De Duve 2005, for example.

8. Lane 2009, 2011, 2015; Lane and Martin 2010. See also note 20.

9. Christian de Duve cites an average cell diameter of 20 microns for eukaryotes versus 1 micron for prokaryotes, which would make the volume difference $20^3 = 8,000$ (2005:187). Actual dimensions vary widely across the groups.

10. Lane 2011, 2015.

11. Lane and Martin 2010.

12. Andrew Bourke (2011) does a wonderful job of discussing these conflicts across scales as a theme of the major transitions of evolution.

13. Blackstone 2013:5.

14. Blackstone 2013, 2016.

15. According to Daniel Lahr and his colleagues, citing Dacks and Roger 1999, "current evidence suggests that sex has a single evolutionary origin and was present in the last common ancestor of eukaryotes" (2011:2081–2082).

16. Bourke 2011 explains the fifty–fifty transfer using the concept of "fair meiosis." Neil Blackstone pointed out to me that "eukaryotic sex is characterized not just by whole-genome recombination but by whole-cell fusion as well" (personal communication, May 2016).

17. Blackstone 2016.

18. Margulis 1971. See Falkowski 2015 for this and subsequent examples of endosymbioses involving the eukaryotic cell. See also Knoll 2015.

19. See Blackstone 2013 and Falkowski 2015.

20. Blackstone 2016 lays out multiple steps needed for the evolution of the eukaryotic cell as cycles of conflict and mediation during the full transition.

21. *Proceedings of the National Academy of Sciences of the United States of America* 112, no. 33 (2015): 10104–10285, has a number of papers discussing and debating issues about the evolution of the eukaryotic cell, and most of the papers are open access.

10. MULTIPLE RAMPS TO THE COMPLEX MULTICELLULAR ORGANISM

1. Grosberg and Strathmann 2007.
2. Volk [1998] 2003.
3. Lane 2015:158.
4. Knoll and Hewitt 2011:253. Knoll and Hewitt say that "complex multicellularity has evolved in animals, plants, fungi (at least three times . . .), green algae, red algae, and brown algae" (253), for a count of eight times. But even if we count fungi only once (some controversy there; Bourke 2011:174 cites two times for fungi), the count is at least six times.
5. John Tyler Bonner (2000, 2006) invokes such a simple, easy beginning to multicellularity at an early stage.
6. Bonner 2006:72–73.
7. Dawkins 1999 is the masterful treatment of this idea.
8. Bourke 2011:14.
9. Eric Brenner, clinical assistant professor at New York University, personal communication, October 2016.
10. See the work David Schwartzman and I have done on the biotic enhancement of weathering (Schwartzman and Volk 1989; Volk [1998] 2003).
11. Rodolfo Llinás (2001:15–17) uses the tunicate to point out the relationship between nervous systems and active, guided movement.
12. On this point, see, for example, King 2004 and other work from Nicole King's lab. However, the nervous systems in ancient, simple animals might have independently evolved twice (Moroz et al. 2014), though this hypothesis is debated.

11. ANIMAL SOCIAL GROUPS WILD WITH POSSIBILITIES

1. I diverge in the definition of this level from the usual next biological transition in the "major transitions of evolution" literature, which tends to focus on eusocial groups (eusocial insects, shrimp, etc.) as having made the "major transition" from multicellularity. For me, a wolf pack is also on this new level of combogenesis. I don't discuss this difference between me and the general literature in more detail, as much as I would like to; it depends on the questions being asked, and I am aiming for consistency within my framework of inquiry, which develops the idea that the non-eusocial but still highly social groups of primates were the subtype within this level that led to the next level and its human tribal metagroups as groups of groups, with culture.
2. Edward O. Wilson, for one, uses this term (2014:92, 165).
3. Bourke 2011:14–15.
4. E. Wilson 2014:20, 31. For example, ants and termites, as two lineages, total thousands of species.
5. Angier 2014.

6. Bonner 2006:72–73.
7. D. Wilson 2007; Bourke 2011; D. Wilson et al. 2014. I cite only two works by David Sloan Wilson here, but please look at his website for more: https://evolution-institute .org/profile/david-sloan-wilson/.
8. Bonner 2000.
9. E. Wilson 2014 cites examples of more complexity in ant societies.
10. For example, studies have found cases involving voles and other small mammals in which styles of social relations can be influenced and thus seem to depend on only a few genes.
11. Andrew Bourke, email communications during his review of a draft of this chapter, June–July 2016; see also Bourke 2011:91, 182, 184.
12. I am indebted to Andrew Bourke for pointing this out to me (email communications, June–July 2016).
13. *Return of the Wolf* 2015.
14. Space prevents me from going into the issue of chemical-based animal borders, such as among ant societies, an issue raised by Andrew Bourke (2011). For a pattern logic to tackle this topic, see Volk 1995 (chapter on borders). On the history of the main types of human "borders," see Diener and Hagan 2012.
15. Aureli et al. 2008.

12. TRIBAL METAGROUPS AND CULTURAL EVOLUTION

1. Dunbar, Gamble, and Gowlett 2014. The "social brain hypothesis" is usually attributed to Robin Dunbar and his large body of work on the topic.
2. See Dyble et al. 2015 on "fluid meta-groups" as relevant to the "selective context for the evolution of human hypercooperation and cumulative culture" (796); see Hill, Walker, et al. 2011 for "metagroups ('tribes')" and "metagroup social structure" (1286).
3. Burkart et al. 2014; Dyble et al. 2015.
4. Pagel 2012a.
5. Dunbar, Gamble, and Gowlett 2014:42.
6. See Hill, Wood, et al. 2014 for both metaband and multiband.
7. Pagel 2012b:passim.
8. John Allen suggested the term *clan* in response to a lecture I gave at Synergia Ranch in New Mexico on July 24, 2014.
9. Hoffecker 2011, 2013. John Hoffecker's Type 1a, an earlier form of "superbrain," could be relevant here as well.
10. Building on work by the philosopher John Searle, Henry Plotkin says: "Remaining true to Searle's position . . . the explanation offered for collective intentionality is that it is an evolved group-selected characteristic of humans. The move from I-intentionality to we-intentionality may mark the crucial transition from the kind of culture chimpanzees have, protoculture, to first human culture, preparatory to its subsequent

shaping to contemporary culture. And culture, of course, is a group-level adaptation" ([2002] 2003:259).

11. See Hill, Walker, et al. 2011; Hill, Wood, et al. 2014.
12. John Hoffecker, personal communication during his reading of a draft of this chap-
    ter, October 2016.
13. Dunbar, Gamble, and Gowlett 2014:40–42.
14. Ibid.:170.
15. From a lecture I heard at the Anthropology Department of New York University.
    Ostrich eggshell beads that are 40,000 years old have been found in sub-Saharan
    Africa (Ambrose 1998).
16. Randall White, personal communication, September, 2016; he approved this paragraph.
17. Boehm 2012.
18. Flannery and Marcus 2012:58.
19. Hill, Wood, et al. 2014.
20. For Steven Pinker (2010), language is one of three major aspects of humans' cognitive
    niche. See also Donald 1991; Sterelny 2015; Szathmáry 2015.
21. Dunbar, Gamble, and Gowlett 2014:42.
22. Mark Pagel (2012a) discusses intervals of 160,000 to 200,000 and 100,000 to 70,000
    years ago for the beginnings of language. Iain Davidson and William Nobel (1992) say
    that full language had to be in place by the time people traversed to greater Australia,
    perhaps by 53,000 years ago, using boats that groups rowed across waters, unable see
    the next land.
23. Cited in Balter 2011:1261.
24. Hodder 2012.
25. See, for example, Suddendorf and Corballis 2007. Endel Tulving was an early key re-
    searcher into the structure and importance of mental time travel.
26. Coward and Gamble 2008:1969.
27. Pinker 2010.
28. Hill, Wood, et al. 2014:1.
29. Kim Hill, personal communication during his reading of a draft of this chapter,
    October 2016. Hill notes the significant modeling work by Joseph Henrich (2004),
    whose "elegant mathematical model . . . makes predictions which have been strongly
    confirmed in subsequent experimental research[,] which clearly confirms that larger
    social networks lead to cumulative cultural evolution. Hence, the metaband struc-
    ture is key to understanding why humans alone have cumulative culture."
30. Hill, Walker, et al. 2011:1286.
31. See, for instance, Mithen 1996 and Hoffecker 2011.
32. Hodder 2012.
33. Kelly 2005.
34. Ridley 2015.
35. Kim Hill, personal communication, October 2016. I am indebted to Hill for outlin-
    ing these connections, which I have put into my own language, among "metaband
    structure [that] can be understood via the human mating patterns, which themselves

probably arise from shifts in the hominid feeding niche." Hill cites Chapais 2008 as important in understanding the ancient innovations in human mating patterns.

36. Referring to the hunter-gatherer groups featured in his research, Hill states: "They call all people who belong to that unit by terms that usually translate as 'person' or 'the people.' This tends to imply that outsiders are not really 'people' in the same sense—I think because accepted social norms do not apply to interactions with outsiders, only with 'the people'" (personal communication, October 13, 2016.) Also, Fred Myers, professor of anthropology at New York University and specialist in Australian aboriginal social systems, told me that in the bands he has studied, a band ("mob" or "camp") of twenty-five to fifty people is just what it happens to be at any moment; he also explained that it is fundamental to organize with people who live elsewhere in space, such as by shared rituals, specialized knowledge, or descent group. Furthermore, it is these larger groupings (with complex categories) that are more enduring through time than the bands themselves. Band membership sometimes can change daily. Crucial is the fact that there are "ways of scaling levels of identity" (personal communication, October 26, 2016).

13. TRANSPLANTABLE AGROVILLAGES

1. On complex hunter-gatherers, see, for example, Hayden 2009. Trevor Watkins says that prior to agriculture there is "evidence of nested networks, local (within communities, among local communities), regional and supra-regional" (2010:631). Also, Tobias Richter and his colleagues show that during the hunter-gatherer period of the late Paleolithic (also called the Epipaleolithic) "long-term and wide-ranging social networks of exchange and interaction existed within and between regions of the southern Levant" (2011:108).

2. Bleed and Matsui 2010.

3. Watkins 2010.

4. Curry 2008; Watkins 2010.

5. Haldorsen et al. 2011.

6. Vigne 2011.

7. On the gradual origin of agriculture, from accidents to conscious control, see Doebley, Gaut, and Smith 2006; Belfer-Cohen and Goring-Morris 2009; Smith 2011.

8. Hodder 2012:197.

9. Smith 2011, 2016.

10. In fact, according to Samuel Bowles (2011), the initial farming systems were less productive than foraging systems, and thus the initial spread of agriculture might have been related to gains in population and social organization. But see the quote from Peter Rowley-Conwy and Robert Layton (2011) in note 13.

11. Smith 2006:fig. 1. Robert Rowthorn and Paul Seabright (2010) list seven to eleven sites. Sylvain Glémin and Thomas Bataillon (2009) cite eight domestication centers for grains (from grasses). Ian Hodder lists nine regions of origin (2012:201).

12. For example, "In Europe, wild oats and rye were initially native weeds that spread into wheat and barley fields, but eventually they did so well that they were taken into cultivation in their own right" (Rowley-Conwy and Layton 2011:857).

13. Rowley-Conwy and Layton 2011:856. Rowley-Conwy and Layton emphasize this point: "In the region where the Near Eastern agricultural system developed, agriculture can support more people per unit area than hunting and gathering. But agriculture as an integrated system closely controlled on a day to day basis by people has a further advantage: it is a niche that can be exported to areas outside the original heartland. Within a few millennia, the Near Eastern system extended from Ireland to northern China (where it encountered and was integrated with the Chinese agricultural system), and from the Urals to the Sudan. No hunter-gatherer niche could match this. The niche was not just transported, it was modified to be able to cope with new environments" (856).

14. Volk 2009b.

15. "The behaviors of things, plants and animals traps humans into various forms of care" (Hodder 2012:85). "The origins of farming and settled life came about in the Middle East as part of a long-term process of entanglement. . . . [T]he things themselves . . . drag humans into dependencies" (Hodder 2012:195).

16. Pollan 2001. According to Peter Bleed and Akira Matsui, "In that sense, agriculture is an ecological niche operated by people, but was 'constructed'—at least partially and initially—by species that interacted with humans. Agriculture grew as successful domesticates directed human effort toward themselves and away from other resources. . . . [S]ome species had qualities that so attracted human investment that people became committed to their survival" (2010:367). Mark Nathan Cohen states that "one species, our own, entered into the coevolutionary interaction with a wide variety of potential domesticates. . . . [H]uman behavior . . . changed in fundamental ways in many regions at roughly the same time and in roughly the same economic contexts" (2009:707, emphasis in original).

14. GEOPOLITICAL STATES, MASTERS OF ACQUISITION AND MERGER

1. Spencer 2014.

2. Flannery and Marcus 2012:555.

3. Ibid.:364.

4. Spencer 2010:7125.

5. Spencer 2014:56, my emphasis.

6. Flannery and Marcus 2012:453. Flannery and Marcus also make the point using their three cases of first-generation kingdoms in the Americas (where Spencer's list has two primary states, a difference that shows some debate going on but does not affect us): "The Zapotec [Mexico], the Moche [Peru], and the Maya [Guatemala] all created monarchies out of rank societies. They did so . . . by forcibly uniting a group of rival societies. . . .

Our three New World monarchies had something else in common. Once having created the apparatus of a kingdom, they expanded against neighboring groups" (392).

7. Redmond and Spencer 2012:22.
8. Spencer 2014:57.
9. Bonner 2006:72–73.
10. Dubreuil 2010:7–8.
11. Ibid.:189–190.
12. Ibid.:213, 229–230.
13. Spencer 2010:7125.
14. Flannery and Marcus 2012:543.
15. Spencer 2014:57.
16. Dubreuil 2010:206–207.
17. Spencer 2010:7121.
18. Dubreuil 2010:206–207.
19. Flannery and Marcus 2012:373. Flannery and Marcus designate the huge, successful empire of Sargon in the Middle East a third-generation state.
20. Solomon, Greenberg, and Pyszczynski 2015. I have summarized this research elsewhere (Volk 2009a).
21. Flannery and Marcus 2012:546–547. Peter Turchin's (2009) research indicates that the waxings and wanings of "mega-empires" (after the primary states) were driven especially by conflicts at the borders of agricultural and nomadic civilizations.
22. Flannery and Marcus 2012:364.

15. DYNAMICAL REALMS AND THEIR BASE LEVELS

1. I ignore here complicated issues about "energy repose" such as the fact that some structural things within a given level from nucleons to molecules have more or less energy repose than others within the same level. For example, nucleons within the atomic nuclei of iron have lower binding energy than nucleons in other atomic nuclei, and an atom of carbon in a methane molecule in the presence of oxygen could have more "repose" if it were in a molecule of carbon dioxide. One can talk about local energy repose. My grouping from fundamental quanta to molecules in what will be called the realm of physical laws is roughly along the lines of what the biochemists Robert Pascal and Addy Pross call "thermodynamic stability" in the "physicochemical world" (2016:509). Also see other work by Pross (2011, 2012, 2013) and by Pascal and Pross (2014).

2. The systems theorist and philosopher Mario Bunge (2003) uses the word *mechanisms*. The humanities systems scholar Bruno Clarke prefers *operations* (personal communication, July 2015). I take these words as alternates equivalent to *dynamics*, as used here.

3. See Pross 2011; Pascal and Pross 2016; and other papers by these authors.

4. Pross 2011:passim.
5. Dobzhansky 1973 (from the title of Dobzhansky's article).
6. For example, I have written on issues about biological evolution and entropy (Volk 2007; Volk and Pauluis 2010).
7. I have elsewhere considered the evolutionary realms as one large group (Volk 2007–2008), in contrast to physics–chemistry, but the emergence of cultural evolution from biological evolution is so significant and the specific dynamics of each so unique and interesting that I consider them two dynamical realms in this study.

16. ALPHAKITS: ATOMIC, GENETIC, LINGUISTIC

1. Close 2004:5.
2. Jordan and Kallenbach have used this phrase several times in various conversations with me.
3. Falkowski 2015:95.
4. In Volk 1995, I used the term *alphabetic holarchy*, which in retrospect is cumbersome. Also, I now refer to "sets" of the alphakit, though "levels" would be fine as well. I think the term *sets* says it better, and in any case I have tried to maintain a specific use of the word *level* in this book.
5. Atkinson 2011.
6. To coin the word *alphakit*, I mutated the word *alphabet*. But for many tens of thousands of years, people were actively working, thinking, and loving without an alphabet and without even nonalphabetic written language. And not all cultures today have an alphabet-based writing. But they all have spoken language, and spoken language has the basic sounds of speech called "phonemes."
7. On recursion, see, for example, Hauser, Chomsky, and Fitch 2002.
8. Hodder 2012.

17. THEMES IN EVOLUTIONARY DYNAMICS

1. Scholars worth looking into include William Calvin (1987, 1997); Gary Cziko (1995); Daniel Dennett (1995); Eva Jablonka and Marion Lamb (2005); Eric Beinhocker (2007); Nathalie Gontier (2007); David Sloan Wilson (2007; also D. Wilson et al. 2014); Joseph Henrich, Robert Boyd, and Peter Richerson (2008); Alex Mesoudi (2009; also Mesoudi, Whiten, and Laland 2006); Chris Buskes (2013, a great place to start if this topic is new to you); Tim Lewens (2013); and Matt Ridley (2015). Full names are given here so you can seek out other work by these people as well as by those scholars they cite and by those who cite them. See also Gregory Bateson on "the two great stochastic processes" (1979:chap. 6).
2. Acerbi and Mesoudi 2015:481.

3. All Darwin quotes are from the famous concluding paragraph of *On the Origin of Species* (Darwin 1859).

4. Calvin 1997.

5. D. Wilson 2007:17220–17221.

6. "Logical skeleton" in Gontier 2007; "Darwin machine" in Calvin 1987, Plotkin 1994, and D. Wilson et al. 2014; "algorithm" in Dennett 1995 and Bostrom 2014:187; "engine" in Mayfield 2013; "formula" in Buskes 2013; "heuristic" in Wynne 2001:352; "recipe" in D. Wilson 2007; and "ingredients" in Calvin 1997.

7. See Pross 2012 as well as other papers by Pross (2011, 2013) and by Pascal and Pross (2014, 2016). Their relevant term is *dynamic-kinetic stability*.

8. Gabora 2013. Liane Gabora establishes terminology and logic for a shift from "communal exchange" to "selectionist dynamics" in the stages of emergence of biological evolution.

9. Discussed, for example, in Cziko 1995 and D. Wilson et al. 2014.

10. A good history of learning in animals is given in D. Wilson et al. 2014 as well as in Burghardt, Stuart, and Shorey 2014.

11. Wynne 2001:355.

12. A large issue. See LeDoux 2015.

13. For example, Donald 1991; Mithen 1996; Hoffecker 2011, 2013; Pagel 2012b; Sterelny 2015. And there are many other outstanding contributions to this topic.

14. Luhmann 2012:132, 135. Luhmann explicitly uses the term *evolution* in his theories of sociology; his own trio of strands consists of (translated from German) "variation," "selection," and restabilization" (see, e.g., 2012:258–259). See also Volk 1995:90 for a discussion of the "yes" and "no" binary as related to other binaries.

15. Szathmáry 2015:10109. Szathmáry is using the term *heredity potential* in its expanded sense, which many scholars in cultural evolution studies do. I would prefer to say "propagation potential." He captures well the exploratory consequences of evolution in his adjective *open-ended*.

16. Boehm 2012.

17. This picture is consistent with Luhmann's (2012) two systems of the psychic (individual) and the social (social) in overall society. It is also consistent with Gabora's (2013) examination of cultural evolution that has a component of individual creativity in an overall communal exchange system of cultural patterns among members of society.

18. Donald 1991.

18. CONVERGENT THEMES OF COMBOGENESIS

1. David Queller (1997 and other works) is often cited as the originator of the distinction (in the study of the major transitions of evolution) between mergings of unlikes ("egalitarian" transitions, of which the origin of the eukaryotic cell is one) and mergings of likes ("fraternal" transitions). Thus, in Queller's terminology, my proposed

parallel involves two "egalitarian" events of combogenesis in different evolutionary realms.

2. I think the term *possibility space* compares favorably to terms others have used for the concept, such as "design hyperspace" (Morris 2009), "adjacent possible" (Kauffman 2008), "design space" (Dennett 1995), and "nonmanifest order" (Salk 1983).
3. Hawking and Mlodinow 2010.
4. Ibid.:160.
5. The question of why there is something rather than nothing has been provocatively covered in Holt 2012 and Gefter 2014.
6. Smolin 1997.
7. Szathmáry 2015:10106.
8. Kauffman 2008.
9. Salk 1983.

EPILOGUE: WHAT ABOUT THE FUTURE?

1. Spencer 2014:57.
2. James Lovelock (2014) uses the date of Thomas Newcomen's steam engine, 1712, as the start of the Anthropocene. An Internet search using the key word *anthropocene* will get loads of results. I have also worked on and written about human-induced changes to the atmosphere (Volk 2008).
3. E. Wilson 2014:117.
4. De Chardin 1959. I have emphasized elsewhere the significance of biogeochemical closure of the biosphere (Barlow and Volk 1990; Volk [1998] 2003).
5. Fuller 1981:xxii.
6. For example, Bostrom 2014. I agree that we need to be collectively thinking much more about negative consequences of artificial intelligence. I also note that my student Halley Young pointed out that our use of electrons changes what those electrons are doing, an example of how human culture incorporates and alters the elements of all the levels that precede it.
7. Salk 1983:11.
8. Einstein 1954:150 (from "A Message to Intellectuals" [1948]).
9. E. Wilson 2014:59–60.
10. Lovelock 2014.
11. Jamieson 2014.
12. Bronowski 1973:435.
13. Salk 1983:120.
14. Einstein 1954:95 (from "The Disarmament Conference of 1931").
15. Ibid.:155 (from the essay "Why Socialism?" [1949]).
16. Bostrom 2014.

17. The concerns about cyborgs and artificial intelligence are explored, for example, in recent movies such as *Her* (Spike Jonze, 2013), *Transcendence* (Wally Pfister, 2014), and *Ex Machina* (Alex Garland, 2015) as well as in novels such as *The Circle* (Dave Eggers, 2013). Transcendent in such exploration are the Olaf Stapledon novels *Last and First Men* (1930) and *Star Maker* (1937).

18. Cziko 1995; Bostrom 2014.

BIBLIOGRAPHY

Acerbi, Alberto, and Alex Mesoudi. 2015. "If We Are All Cultural Darwinians What's the Fuss About? Clarifying Recent Disagreements in the Field of Cultural Evolution." *Biology and Philosophy* 30:481–503. doi:10.1007/s10539-015-9490-2.

Ambrose, Stanley H. 1998. "Chronology of the Later Stone Age and Food Production in East Africa." *Journal of Archaeological Science* 25:377–392.

Angier, Natalie. 2014. "When Trilobites Ruled the World." *New York Times*, March 3.

Atkinson, Quentin D. 2011. "Phonemic Diversity Supports a Serial Founder Effect Model of Language Expansion from Africa." *Science* 332:346–349.

Aureli, Filippo, Colleen M. Schaffner, Christophe Boesch, Simon K. Bearder, Josep Call, Colin A. Chapman, Richard Connor, et al. 2008. "Fission–Fusion Dynamics: New Research Frameworks." *Current Anthropology* 49:627–654.

Balter, Michael. 2011. "South African Cave Slowly Shares Secrets of Human Culture." *Science* 332:1260–1261.

Barabási, Albert-Laszlo. 2003. *Linked: How Everything Is Connected to Everything Else and What It Means for Business, Science, and Everyday Life.* New York: Plume.

Barlow, Connie, and Tyler Volk. 1990. "Open Living Systems in a Closed Biosphere: A New Paradox for the Gaia Debate." *BioSystems* 23:371–384.

Bateson, Gregory. 1979. *Mind and Nature: A Necessary Unity.* New York: Dutton.

Beinhocker, Eric D. 2007. *The Origin of Wealth: The Radical Remaking of Economics and What It Means for Business and Society.* Cambridge, Mass.: Harvard Business Review Press.

Belfer-Cohen, Anna, and Nigel Goring-Morris. 2009. "For the First Time." *Current Anthropology* 50:669–672.

Bertalanffy, Ludwig von. [1968] 2015. *General System Theory: Foundations, Development, Applications.* Reprint. New York: George Braziller.

Bianconi, E., A. Piovesan, F. Facchin, A. Beraudi, R. Casadei, F. Frabetti, L. Vitale, et al. 2013. "An Estimation of the Number of Cells in the Human Body." *Annals of Human Biology* 40:463–471.

Blackstone, Neil W. 2013. "Why Did Eukaryotes Evolve Only Once? Genetic and Energetic Aspects of Conflict and Conflict Mediation." *Philosophical Transactions of the Royal Society B* 368, no. 1622. http://dx.doi.org/:10.1098/rstb.2012.0266.

——. 2016. "An Evolutionary Framework for Understanding the Origin of Eukaryotes." *Biology* 5, no. 2. http://dx.doi.org/10.3390/biology5020018.

Bleed, Peter, and Akira Matsui. 2010. "Why Didn't Agriculture Develop in Japan? A Consideration of Jomon Ecological Style, Niche Construction, and the Origins of Domestication." *Journal of Archaeological Method and Theory* 17:356–370. doi: 10.1007/s10816-010-9094-8.

Boehm, Christopher. 2012. *Moral Origins: The Evolution of Virtue, Altruism, and Shame.* New York: Basic.

Bonner, John Tyler. 2000. *First Signals: The Evolution of Multicellular Development.* Princeton, N.J.: Princeton University Press.

——. 2006. *Why Size Matters: From Bacteria to Blue Whales.* Princeton, N.J.: Princeton University Press.

Bostrom, Nick. 2014. *Superintelligence: Paths, Dangers, Strategies.* Oxford: Oxford University Press.

Bourke, Andrew F. G. 2011. *Principles of Social Evolution.* Oxford: Oxford University Press.

Bowles, Samuel. 2011. "Cultivation of Cereals by the First Farmers Was Not More Productive Than Foraging." *Proceedings of the National Academy of Sciences of the United States of America* 108:4760–4765.

Bronowski, Jacob. 1973. *The Ascent of Man.* New York: Little, Brown.

Bunge, Mario. 2003. *Emergency and Convergence: Qualitative Novelty and the Unity of Knowledge.* Toronto: University of Toronto Press.

Burghardt, Gordon M., Gregory L. Stuart, and Ryan C. Shorey. 2014. "Why Can't We All Just Get Along? Integration Needs More Than Stories." *Behavioral and Brain Sciences* 37:420–421. doi: 10.1017/S0140525X13001593.

Burkart, J. M., O. Allon, F. Amici, C. Fichtel, C. Finkenwirth, A. Heschl, J. Huber, et al. 2014. "The Evolutionary Origin of Human Hyper-cooperation." *Nature Communications* 5:article no. 4747. doi: 10.1038/ncomms5747.

Buskes, Chris. 2013. "Darwinism Extended: A Survey of How the Idea of Cultural Evolution Evolved." *Philosophia* 41:661–691. doi: 10.1007/s11406-013-9415-8.

Calcott, Brett, and Kim Sterelny, eds. 2011. *The Major Transitions in Evolution Revisited.* Cambridge, Mass.: MIT Press.

Calvin, William H. 1987. "The Brain as a Darwin Machine." *Nature* 330:33–34.

——. 1997. "The Six Essentials? Minimal Requirements for the Darwinian Bootstrapping of Quality." *Journal of Memetics* 1, no. 1. http://www.williamcalvin.com /1990s/1997JMemetics.htm.

Chapais, Bernard. 2008. *Primeval Kinship: How Pair-Bonding Gave Birth to Human Society.* Cambridge, Mass.: Harvard University Press.

Close, Frank. 2004. *Particle Physics: A Very Short Introduction*. Oxford: Oxford University Press.

Cohen, Mark Nathan. 2009. Comment in "Replies." In "Re-thinking the Origins of Agriculture," special issue of *Current Anthropology* 50:707.

Corning, Peter A., and Eörs Szathmáry. 2015. "'Synergistic Selection': A Darwinian Frame for the Evolution of Complexity." *Journal of Theoretical Biology* 371:45–58.

Coward, Fiona, and Clive Gamble. 2008. "Big Brains, Small Worlds: Material Culture and the Evolution of Mind." *Philosophical Transactions of the Royal Society B* 363:1969–1979. doi: 10.1098/rstb.2008.0004.

Curry, Andrew. 2008. "Seeking the Roots of Ritual." *Science* 319:278–280.

Cziko, Gary. 1995. *Without Miracles: Universal Selection Theory and the Second Darwinian Revolution*. Cambridge, Mass.: MIT Press.

Dacks, J., and A. J. Roger. 1999. "The First Sexual Lineage and the Relevance of Facultative Sex." *Journal of Molecular Evolution* 48:779–783. doi: 10.1007/PL00013156.

Darling, David. n.d. "Interstellar Molecules." In *The Encyclopedia of Science*. http://www.daviddarling.info/encyclopedia/I/ismols.html.

Darwin, Charles. 1859. *On the Origin of Species by Means of Natural Selection, or the Preservation of Favoured Races in the Struggle for Life*. London: Murray. http://darwin-online.org.uk.

Davidson, Iain, and William Noble. 1992. "Why the First Colonisation of the Australian Region Is the Earliest Evidence of Modern Human Behaviour." *Archaeology in Oceania* 27:135–142.

Dawkins, Richard. 1999. *The Extended Phenotype*. Rev. ed. Oxford: Oxford University Press.

De Chardin, Pierre Teilhard. 1959. *The Phenomenon of Man*. New York: Harper. A translation of *Le phénomène humain*. Paris: Éditions du Seuil, 1955.

De Duve, Christian. 2005. *Singularities: Landmarks on the Pathways of Life*. Cambridge: Cambridge University Press.

Dennett, Daniel. 1995. *Darwin's Dangerous Idea: Evolution and the Meanings of Life*. New York: Simon and Schuster.

Diener, Alexander C., and Joshua Hagan. 2012. *Borders: A Very Short Introduction*. New York: Oxford University Press.

Dobzhansky, Theodosius. 1973. "Nothing in Biology Makes Sense Except in the Light of Evolution." *American Biology Teacher* 35:125–129.

Doebley, John F., Brandon S. Gaut, and Bruce D. Smith. 2006. "The Molecular Genetics of Crop Domestication." *Cell* 127:1309–1321.

Donald, Merlin. 1991. *Origins of the Modern Mind: Three Stages in the Evolution of Culture and Cognition*. Cambridge, Mass.: Harvard University Press.

Donaldson, Mike. 2012. "The Gwion or Bradshaw Art Style of Australia's Kimberley Region Is Undoubtedly Among the Earliest Rock Art in the Country—but Is It Pleistocene?" In Jean Clottes, dir., *Pleistocene Art of the World, Symposium 5: Pleistocene*

Art in Australia, 1001–1012. http://blogs.univ-tlse2.fr/palethnologie/wp-content/files/2013/fr-FR/version-longue/articles/AUS3_Donaldson.pdf.

Dryden, David T. F., Andrew R. Thomson, and John H. White. 2008. "How Much of Protein Sequence Space Has Been Explored by Life on Earth?" *Journal of the Royal Society Interface* 5:953–956.

Dubreuil, Benoît. 2010. *Human Evolution and the Origins of Hierarchies: The State of Nature.* Cambridge: Cambridge University Press.

Dunbar, Robin, Clive Gamble, and John Gowlett. 2014. *Thinking Big: How the Evolution of Social Life Shaped the Human Mind.* London: Thames and Hudson.

Dyble, A., G. D. Salali, N. Chaudhary, A. Page, D. Smith, J. Thompson, L. Vinicius, et al. 2015. "Sex Equality Can Explain the Unique Social Structure of Hunter-Gatherer Bands." *Science* 348:796–798. doi: 10.1126/science.aaa5139.

Einstein, Albert. 1954. *Ideas and Opinions.* Translated by Sonja Bargmann. New York: Bonanza.

Falkowski, Paul. 2015. *Life's Engines: How Microbes Made Earth Habitable.* Princeton, N.J.: Princeton University Press.

Flannery, Kent, and Joyce Marcus. 2012. *The Creation of Inequality: How Our Prehistoric Ancestors Set the Stage for Monarchy, Slavery, and Empire.* Cambridge, Mass.: Harvard University Press.

Fuller, Buckminster. 1981. *Critical Path.* New York: St. Martin's Press.

Gabora, Liane. 2013. "An Evolutionary Framework for Cultural Change: Selectionism Versus Communal Exchange." *Physics of Life Reviews* 10:117–145.

Gefter, Amanda. 2014. *Trespassing on Einstein's Lawn: A Father, a Daughter, the Meaning of Nothing, and the Beginning of Everything.* New York: Bantam.

Glémin, Sylvain, and Thomas Bataillon. 2009. "A Comparative View of the Evolution of Grasses Under Domestication." *New Phytologist* 183:273–290. doi: 10.1111/j.1469-8137.2009.02884.x.

Gontier, Nathalie. 2007. "Universal Symbiogenesis: An Alternative to Universal Selectionist Accounts of Evolution." *Symbiosis* 44:167–181.

Gribbin, John. 2013. *Erwin Schrödinger and the Quantum Revolution.* Hoboken, N.J.: Wiley.

Grosberg, Richard K., and Richard R. Strathmann. 2007. "The Evolution of Multicellularity: A Minor Major Transition?" *Annual Review of Ecology, Evolution, and Systematics* 38:621–654.

Haldorsen, Sylvi, Hasan Akan, Bahatin Çelik, and Manfred Heun. 2011. "The Climate of the Younger Dryas as a Boundary for Einkorn Domestication." *Vegetation History and Archaeobotany* 20:305–318. doi: 10.1007/s00334-011-0291-5.

Hauser, Marc D., Noam Chomsky, and W. Tecumseh Fitch. 2002. "The Faculty of Language: What Is It, Who Has It, and How Did It Evolve?" *Science* 298:1569–1579.

Hawking, Stephen, and Leonard Mlodinow. 2010. *The Grand Design.* New York: Bantam.

Hayden, Brian. 2009. Comment in "Replies." In "Re-thinking the Origins of Agriculture," special issue of *Current Anthropology* 50:708–709.

Hazen, Robert M. 2009. "The Emergence of Patterning in Life's Origin and Evolution." *International Journal of Developmental Biology* 53:683–692.

Hazen, R. M., D. Papineau, W. Bleeker, R. T. Downs, J. Ferry, T. McCoy, D. Sverjensky, and H. Yang. 2008. "Mineral Evolution." *American Mineralogist* 93:1693–1720.

Henrich, Joseph. 2004. "Demography and Cultural Evolution: How Adaptive Cultural Processes Can Produce Maladaptive Losses: The Tasmanian Case." *American Antiquity* 69:197–214.

Henrich, Joseph, Robert Boyd, and Peter J. Richerson. 2008. "Five Misunderstandings About Cultural Evolution." *Human Nature* 19, no. 2: 119–137. doi: 10.1007/s12110 -008-9037-1.

Hill, Kim R., Robert S. Walker, Miran Božičević, James Eder, Thomas Headland, Barry Hewlett, A. Magdalena Hurtado, et al. 2011. "Co-residence Patterns in Hunter-Gatherer Societies Show Unique Human Social Structure." *Science* 331:1286–1289. doi: 10.1126/science.1199071.

Hill, Kim R., Brian M. Wood, Jacopo Baggio, A. Magdalena Hurtado, and Robert T. Boyd. 2014. "Hunter-Gatherer Inter-band Interaction Rates: Implications for Cumulative Culture." *PLOS One* 9, no. 7: 9 pages. e102806 (open access).

Hodder, Ian. 2012. *Entangled: An Archaeology of the Relationships Between Humans and Things.* Chichester, U.K.: Wiley-Blackwell.

Hoffecker, John F. 2011. *Landscape of the Mind: Human Evolution and the Archeology of Thought.* New York: Columbia University Press.

——. 2013. "The Information Animal and the Super-brain." *Journal of Archaeological Method and Theory* 20:18–41.

Holt, Jim. 2012. *Why Does the World Exist? An Existential Detective Story.* New York: Liveright.

Jablonka, Eva, and Marion J. Lamb. 2005. *Evolution in Four Dimensions: Genetic, Epigenetic, Behavioral, and Symbolic Variation in the History of Life.* Cambridge, Mass.: MIT Press.

Jamieson, Dale. 2014. *Reason in a Dark Time: Why the Struggle Against Climate Change Failed—and What It Means for Our Future.* Oxford: Oxford University Press.

Johnson, Steven. 2001. *Emergence: The Connected Lives of Ants, Brains, Cities, and Software.* New York: Scribner.

Kauffman, Stuart A. 2008. *Reinventing the Sacred: A New View of Science, Reason, and Religion.* New York: Basic.

Kelly, Robert L. 2005. "Hunter-Gatherers, Archeology, and the Role of Selection in the Evolution of the Human Mind." In *A Catalyst for Ideas: Anthropological Archeology and the Legacy of Douglas W. Schwartz*, edited by Vernon Scarborough and Richard Leventhal, 19–39. Santa Fe: School of American Research Press.

King, Nicole. 2004. "The Unicellular Ancestry of Animal Development." *Developmental Cell* 7:313–325.

Knoll, Andrew H. 2015. *Life on a Young Planet: The First Three Billion Years of Evolution on Earth.* Updated ed. Princeton, N.J.: Princeton University Press.

Knoll, Andrew H., and David Hewitt. 2011. "Phylogenetic, Functional, and Geological Perspectives on Complex Multicellularity." In *The Major Transitions in Evolution Revisited*, edited by Brett Calcott and Kim Sterelny, 251–270. Cambridge, Mass.: MIT Press.

Kronfeld, Andreas S. 2008. "The Weight of the World Is Quantum Chromodynamics." *Science* 322:1198–1199.

Lahr, Daniel J. G., Laura Wegener Parfrey, Edward A. D. Mitchell, and Enrique Lara. 2011. "The Chastity of Amoebae: Re-evaluating Evidence for Sex in Amoeboid Organisms." *Proceedings of the Royal Society B* 278:2081–2090. doi: 10.1098/rspb .2011.0289.

Lane, Nick. 2009. *Life Ascending: The Ten Great Inventions of Evolution*. New York: Norton.

——. 2011. "Energetics and Genetics Across the Prokaryote–Eukaryote Divide." *Biology Direct* 6:35. http://biologydirect.biomedcentral.com/articles/10.1186/1745-6150 -6-35.

——. 2015. *The Vital Question: Energy, Evolution, and the Origins of Complex Life*. New York: Norton.

Lane, Nick, and William Martin. 2010. "The Energetics of Genome Complexity." *Nature* 467:929–934.

LeDoux, Joseph. 2015. *Anxious: Using the Brain to Understand and Treat Fear and Anxiety*. New York: Viking.

Lewens, Tim. 2013. "Cultural Evolution." In *The Stanford Encyclopedia of Philosophy*, edited by Edward N. Zalta. Stanford: Stanford University Press. http://plato.stan ford.edu/archives/spr2013/entries/evolution-cultural/.

Llinás, Rodolfo R. 2001. *I of the Vortex: From Neurons to Self*. Cambridge, Mass.: MIT Press.

Lovelock, James. 2014. *A Rough Ride to the Future*. New York: Overlook.

Lucretius. 1994. *On the Nature of the Universe*. Translated by Ronald E. Latham. New York: Penguin.

——. 2010. *The Nature of Things*. Translated by Ian Johnson. Arlington, Va.: Richer Resources.

Luhmann, Niklas. 2012. *Theory of Society*. Vol. 1. Translated by Rhodes Barrett. Stanford: Stanford University Press.

Margulis, Lynn. 1971. *Origin of Eukaryotic Cells*. New Haven, Conn.: Yale University Press.

Martin, William F. 2011. "Early Evolution Without a Tree of Life." *Biology Direct* 6:e36. http://www.biology-direct.com/content/6/1/36.

Mayfield, John E. 2013. *The Engine of Complexity: Evolution as Computation*. New York: Columbia University Press.

McShea, Daniel W., and Robert N. Brandon. 2010. *Biology's First Law: The Tendency for Diversity and Complexity to Increase in Evolutionary Systems*. Chicago: Chicago University Press.

McShea, Daniel W., and Carl Simpson. 2011. "The Miscellaneous Transitions in Evolution." In *The Major Transitions in Evolution Revisited*, edited by Brett Calcott and Kim Sterelny, 19–32. Cambridge, Mass.: MIT Press.

Mesoudi, Alex. 2009. "How Cultural Evolutionary Theory Can Inform Social Psychology and Vice Versa." *Psychological Review* 116:929 952.

Mesoudi, Alex, Andrew Whiten, and Kevin N. Laland. 2006. "Towards a Unified Science of Cultural Evolution." *Behavioral and Brain Sciences* 29:329–347.

Mithen, Steven. 1996. *Prehistory of the Mind: The Cognitive Origins of Art, Religion, and Science*. London: Thames and Hudson.

Morowitz, Harold J. 2002. *The Emergence of Everything: How the World Became Complex*. Oxford: Oxford University Press.

Moroz, Leonid L., Kevin M. Kocot, Mathew R. Citarella, Sohn Dosung, Tigran P. Norekian, Inna S. Povolotskaya, Anastasia P. Grigorenko, et al. 2014. "The Ctenophore Genome and the Evolutionary Origins of Neural Systems." *Nature* 510:109–114.

Morris, Simon Conway. 2009. "The Predictability of Evolution: Glimpses into a Post-Darwinian World." *Naturwissenschaften* 96:1313–1337. doi: 10.1007/s00114-009-0607-9.

O'Malley, Maureen A., and Russell Powell. 2016. "Major Problems in Evolutionary Transitions: How a Metabolic Perspective Can Enrich Our Understanding of Macroevolution." *Biology and Philosophy* 31:159–189.

Pagel, Mark. 2012a. "Evolution: Adapted to Culture." *Nature* 482:297–299.

——. 2012b. *Wired for Culture*. New York: Norton.

Pascal, Robert, and Addy Pross. 2014. "The Nature and Mathematical Basis for Material Stability in the Chemical and Biological Worlds." *Journal of Systems Chemistry* 5, no. 3: 8 pages. doi: 10.1186/1759-2208-5-3.

——. 2016. "The Logic of Life." *Origins of Life and Evolution of Biospheres* 46:507–513.

Pinker, Steven. 2010. "The Cognitive Niche: Coevolution of Intelligence, Sociality, and Language." *Proceedings of the National Academy of Sciences of the United States of America* 107:8993–8999.

Plotkin, Henry. 1994. *Darwin Machines and the Nature of Knowledge*. Cambridge, Mass.: Harvard University Press.

——. [2002] 2003. *The Imagined World Made Real: Towards a Natural Science of Culture*. Paperback ed. New Brunswick, N.J.: Rutgers University Press.

Pollan, Michael. 2001. *The Botany of Desire: A Plant's-Eye View of the World*. New York: Random House.

Pross, Addy. 2011. "Toward a General Theory of Evolution: Extending Darwinian Theory to Inanimate Matter." *Journal of Systems Chemistry* 2:1–14.

——. 2012. *What Is Life? How Chemistry Becomes Biology*. Oxford: Oxford University Press.

——. 2013. "The Evolutionary Origin of Biological Function and Complexity." *Journal of Molecular Evolution* 76:185–191.

Queller, David C. 1997. "Cooperators Since Life Began." *Quarterly Review of Biology* 72:184–188.

Randall, Lisa. 2011. *Knocking on Heaven's Door: How Physics and Scientific Thinking Illuminate the Universe and the Modern World.* New York: Ecco.

Redmond, Elsa M., and Charles S. Spencer. 2012. "Chiefdoms at the Threshold: The Competitive Origins of the Primary State." *Journal of Anthropological Archaeology* 31:22–37.

Reinhard, Rachel, Irith Wyschkony, Sabine Riehl, and Harald Huber. 2002. "The Ultrastructure of *Ignicoccus*: Evidence for a Novel Outer Membrane and for Intracellular Vesicle Budding in an Archaeon." *Archaea* 1:9–18.

Return of the Wolf. 2015. Journeys with Wildlife video. *National Geographic*, July 20. https://www.youtube.com/watch?v=IM5PdNSPeZc.

Reynolds, Gretchen. 2015. "An Upbeat Emotion That's Surprisingly Good for You." *New York Times Magazine*, March 29.

Richter, Tobias, Andrew N. Garrard, Samantha Allock, and Lisa A. Maher. 2011. "Interaction Before Agriculture: Exchanging Material and Sharing Knowledge in the Final Pleistocene Levant." *Cambridge Archaeological Journal* 21:95–114.

Riddihough, Guy. 2015. "Exploring the Limits of Protein Sequence Space." *Science Signaling* 8:ec32. http://stke.sciencemag.org/content/8/363/ec32.

Ridley, Matt. 2015. *The Evolution of Everything: How New Ideas Emerge.* New York: Harper.

Rowley-Conwy, Peter, and Robert Layton. 2011. "Foraging and Farming as Niche Construction: Stable and Unstable Adaptations." *Philosophical Transactions of the Royal Society B* 366:849–862. doi: 10.1098/rstb.2010.0307.

Rowthorn, Robert, and Paul Seabright. 2010. *Property Rights, Warfare, and the Neolithic Transition.* Working Paper no. 10-207. Toulouse: Toulouse School of Economics. http://publications.ut-capitole.fr/3534/.

Salk, Jonas. 1983. *Anatomy of Reality: Merging of Intuition and Reason.* New York: Columbia University Press.

——. 1985. "The Next Evolutionary Step in the Ascent of Man in the Cosmos." *Leonardo* 41, no. 3: 281–286.

Schumm, Bruce A. 2004. *Deep Down Things: The Breathtaking Beauty of Particle Physics.* Baltimore: Johns Hopkins University Press.

Schwartzman, D. W., and T. Volk. 1989. "Biotic Enhancement of Weathering and the Habitability of Earth." *Nature* 340:457–460.

Smith, Bruce D. 2006. "Eastern North America as an Independent Center of Plant Domestication." *Proceedings of the National Academy of Sciences of the United States of America* 103:12223–12228.

——. 2011. "General Patterns of Niche Construction and the Management of 'Wild' Plant and Animal Resources by Small-Scale Pre-industrial Societies." *Philosophical Transactions of the Royal Society B* 366:836–848. doi: 10.1098/rstb.2010.0253.

——. 2016. "Neo-Darwinism, Niche Construction Theory, and the Initial Domestication of Plants and Animals." *Evolutionary Ecology* 30:307–324.

Smith, John Maynard, and Eörs Szathmáry. 1998. *The Major Transitions in Evolution.* Oxford: Oxford University Press.

Smolin, Lee. 1997. *The Life of the Cosmos.* Oxford: Oxford University Press.

Solomon, Sheldon, Jeff Greenberg, and Tom Pyszczynski. 2015. *The Worm at the Core: On the Role of Death in Life.* New York: Random House.

Spencer, Charles S. 2010. "Territorial Expansion and Primary State Formation." *Proceedings of the National Academy of Sciences of the United States of America* 107:7119–7126.

——. 2014. "Modeling the Evolution of Bureaucracy: Political-Economic Reach and Administrative Complexity." *Social Evolution and History* 13:42–66.

Sterelny, Kim. 2015. "Deacon's Challenge: From Calls to Words." *Topoi*, March 11. doi: 10.1007/s11245-014-9284-1.

Strassler, Matt. n.d.a. "Protons and Neutrons: The Massive Pandemonium in Matter." *Of Particular Significance.* https://profmattstrassler.com/articles-and-posts/particle-physics-basics/the-structure-of-matter/protons-and-neutrons/.

——. n.d.b. "What's a Proton, Anyway?" *Of Particular Significance.* https://profmattstrassler.com/articles-and-posts/largehadroncolliderfaq/whats-a-proton-anyway/.

Suddendorf, Thomas, and Michael C. Corballis. 2007. "The Evolution of Foresight: What Is Mental Time Travel, and Is It Unique to Humans?" *Behavioral and Brain Sciences* 30:299–313.

Szathmáry, Eörs. 2015. "Toward Major Evolutionary Transitions Theory 2.0." *Proceedings of the National Academy of Sciences of the United States of America* 112, no. 33: 10104–10111. doi: 10.1073/pnas.1421398112.

Tegmark, Max. 2014. *Our Mathematical Universe: My Quest for the Ultimate Nature of Reality.* New York: Knopf.

Turchin, Peter. 2009. "A Theory for Formation of Large Empires." *Journal of Global History* 4:191–217. doi: 10.1017/S174002280900312X.

Vayenas, Constantinos G., and Stamatios N.-A. Souentie. 2012. *Gravity, Special Relativity, and the Strong Force.* New York: Springer.

Veltman, Martinus J. G. 2003. *Facts and Mysteries in Elementary Particle Physics.* London: World Scientific.

Vigne, Jean-Denis. 2011. "The Origins of Animal Domestication and Husbandry: A Major Change in the History of Humanity and the Biosphere." *Comptes Rendus Biologies* 334:171–181.

Volk, Tyler. 1995. *Metapatterns Across Space, Time, and Mind.* New York: Columbia University Press.

——. 2002. "Toward a Future for Gaia Theory." *Climatic Change* 52:423–430.

——. [1998] 2003. *Gaia's Body: Toward a Physiology of the Earth.* Paperback ed. Cambridge, Mass.: MIT Press, 2003.

——. 2007. "The Properties of Organisms Are Not Tunable Parameters Selected Because They Create Maximum Entropy Production on the Biosphere Scale: A By-Product Framework in Response to Kleidon." *Climatic Change* 85:251–258.

——. 2007–2008. "What Is a Sphere? Metapatterns and Scale-Transcending Functional Principles." *General Semantics Bulletin* 74–75:69–72.

——. 2008. *CO₂ Rising: The World's Greatest Environmental Challenge.* Cambridge, Mass.: MIT Press.

——. 2009a. *Death.* In Tyler Volk and Dorion Sagan, *Death & Sex*, 1–105. White River Junction, Vt.: Chelsea Green.

——. 2009b. "Thermodynamics and Civilization: From Ancient Rivers to Fossil Fuel Energy Servants." *Climatic Change* 95:433–438.

Volk, Tyler, and Jeffry W. Bloom. 2007. "The Use of Metapatterns for Research into Complex Systems of Teaching, Learning, and Schooling, Part I: Metapatterns in Nature and Culture." *Complicity* 4:25–43.

Volk, Tyler, Jeffry W. Bloom, and John Richards. 2007. "Toward a Science of Metapatterns: Building Upon Bateson's Foundation." *Kybernetes* 36:1070–1080.

Volk, Tyler, and Olivier Pauluis. 2010. "It's Not the Entropy You Produce, Rather, How You Produce It." *Philosophical Transactions of the Royal Society B* 365:1317–1322.

Watkins, Trevor. 2010. "New Light on Neolithic Revolution in South-west Asia." *Antiquity* 84:621–634.

Weinberg, Steven. 1977. *The First Three Minutes: A Modern View of the Origin of the Universe.* New York: Basic Books.

Wilczek, Frank. 2008. *The Lightness of Being: Mass, Ether, and the Unification of Forces.* New York: Basic Books.

Wilson, David Sloan. 2007. *Evolution for Everyone: How Darwin's Theory Can Change the Way We Think About Our Lives.* New York: Bantam Dell.

Wilson, David Sloan, Steven C. Hayes, Anthony Biglan, and Dennis D. Embry. 2014. "Evolving the Future: Toward a Science of Intentional Change." *Behavioral and Brain Sciences* 37:395–460. doi: 10.1017/S0140525X13001593.

Wilson, Edward O. 1998. *Consilience: The Unity of Knowledge.* New York: Knopf.

——. 2014. *The Meaning of Human Existence.* New York: Liveright.

Wittgenstein, Ludwig. *Tractatus logico-philosophicus.* New York: Harcourt Brace, 1922.

Woese, Carl R., and George E. Fox. 1977. "Phylogenetic Structure of the Prokaryotic Domain: The Primary Kingdoms." *Proceedings of the National Academy of Sciences of the United States of America* 74:5088–5090.

Woese, Carl R., Otto Kandler, and Mark L. Wheelis. 1990. "Towards a Natural System of Organisms: Proposal for the Domains Archaea, Bacteria, and Eucarya." *Proceedings of the National Academy of Sciences of the United States of America* 87:4576–4579.

Wynne, Clive D. L. 2001. "Universal Plotkinism: A Review of Henry Plotkin's *Darwin Machines and the Nature of Knowledge.*" *Journal of the Experimental Analysis of Behavior* 76:351–361. doi: 10.1901/jeab.2001.76-351. http://www.ncbi.nlm.nih.gov /pmc/articles/PMC1284844/pdf.

INDEX

Acerbi, Alberto, 169

Ache, 117, 119

agriculture, 130–36, 186, 187, 223n10, 224n13, 224n16. *See also* agrovillages

agrovillages, 22, 127–36; and cultural evolution, 150; new relations, 134–36; origins, 127–34, 185; and origins of the geopolitical state, 140, 144; parallels with eukaryotic cells, 183–88; and transplantability of life-support systems, 134–36, 144

algae, 62–63, 80, 88, 92, 96, 99, 220n4

alphakits, 157–65, 226n6; atomic alphakit/ alphabet, 30, 49, 55, 60, 64–65, 157–60, 162, 174–76, 180, 181, 190; and biological evolution, 157, 162–65, 171–73, 175–77; and chemical evolution, 174–76; and computer code, 202–3; and convergent combogenesis, 192; cornucopia sets, 65, 158, 161–65, 168, 172; defined, 65, 160–61; element sets, 65, 158, 160–62, 168, 172; and evolutionary dynamics, 167–69, 171–82, 184, 190; "field of possibilities" set, 162, 172, 191; and fundamental quanta, 30, 157–60, 162; genetic alphakit, 77, 157, 162–65, 167–68, 175, 192; linguistic alphakit, 163–65, 167–68, 178–79, 192; parallel between biological and cultural

alphakits, 165; and physical laws, 30, 157–60, 162

amino acids, 61, 70–71, 76, 77, 163, 172

amoebas, 87, 88, 94, 98

animals: and biological evolution, 150; bounded bodies of, 106, 128; circulatory systems, 95; and cognitive evolutionary dynamics, 176–77; domestication of, 127, 130–36, 154, 186 (*see also* agrovillages); and going back in time by going smaller in scale, 7; and learning, 98; as members of single lineage, 99; new relations, 96–99; origins, 7, 13, 91–92, 96–99 (*see also under* multicellular organisms); remote sensing/nervous systems, 97–99; subclasses, 14. *See also* multicellular organisms

animal social groups, 22, 101–11, 180–82, 220n1; advantages and disadvantages, 104–5, 107; and biological evolution, 152; and cognitive evolutionary dynamics, 176–77; eusocial animals, 102–3, 105, 106, 220n1; fission–fusion societies, 110–11, 121; and genes, 106–7; in-groups and out-groups (social boundaries), 108–11; lack of physical boundaries, 152; new relations, 108–10; origins, 103–5; simplicity relative to multicellular organisms, 105–8; social learning,